The London Letters of Samuel Molyneux, 1712–13

The London Letters of Samuel Molyneux, 1712–13

with an Introduction and Commentary
by
Paul Holden

Edited by
Ann Saunders

with an Epilogue
by
Sheila O'Connell

Publication No. 172
London Topographical Society
2011

3 Meadway Gate
London NW11 7LA
2011

ISBN
978 0 902087 59 0

PRODUCED IN GREAT BRITAIN BY
OUTSET SERVICES LIMITED

CONTENTS

In memory of my father,
William Holden (1923–2011)

EDITOR'S PREFACE

With this publication, your Council has stepped beyond its usual limits, which are the boundaries of London at any particular period. In two of these seven letters, the writer Samuel Molyneux ventures as far as Oxford and Cambridge. Your editor felt that to omit the two excursions would be like pulling the currants out of a bun and throwing them away, leaving the whole mangled; she hopes you will agree.

We have tried, by means of a marked version of Rocque's 1746 *Plan of the Cities of London and Westminster*, to show how small and compact early eighteenth-century London was; from his lodgings near to the Royal Mews — today's Trafalgar Square — Samuel was little more than half an hour's brisk walk from any of his destinations. The collectors, scholars, scientists, engravers, booksellers and craftsmen would, for the most part, have been familiar to each other, at least by name.

These letters provide an archaeology of our national collections and link Queen Anne's reign to the twenty-first century.

Molyneux's career, which had begun so brightly, ended abruptly in ridicule and early death. In her Epilogue, Sheila O'Connell shows the fragility of the early Enlightenment and its vulnerability to entrenched popular beliefs.

We hope that you will enjoy this publication, even if it is not strictly topographical and escapes the boundaries of London.

Acknowledgements

Paul Holden recognised the importance of the Molyneux manuscript and has expressed his thanks to many people and institutions; I join him in grateful acknowledgement. I also wish to thank the staff of Richmond Public Library who were particularly generous with their time, as was the Revd Nigel Worn of St Anne's, Kew Green. The London Library was, as always, a support. Dr Alison Morrison-Low has guided me with Molyneux's scientific work, Vivienne Aldous has solved palaeographical problems and Sheila O'Connell has given an account of the caricatures that resulted from Molyneux's unfortunate mistake. Dr Lesley Miller, Professor Trevor Lloyd, Peter Clayton, Colin Thom and Jeremy Smith also gave valuable help. The Society is deeply grateful to the National Trust, the British Museum, the National Portrait Gallery, and Sampson Lowe for generous help with the illustrations. Bruce Saunders has worked patiently on the map of

Molyneux's travels within London and Graham Maney of Outset Services has toiled to make an awkward manuscript into what we hope is a clear reality. Stephen Croad has kept an eye on the text throughout the development; Roger Cline has manfully compiled the index.

ANN SAUNDERS

FOREWORD

In the eighteenth century, London, in the wider sense of the City and Westminster, became the second largest city in Europe (after Naples). It became the capital of one of, if not the most important of, the major world political and economic powers, with a burgeoning world-wide empire. So why are topographical descriptions of this city so relatively rare, particularly for the start of the century? (The same rarity can also be found amongst political diaries for the same period and, indeed, those topographical descriptions which have been published are usually in the nature of diaries.) Thus the discovery of a series of letters written in 1712–13 by Samuel Molyneux, which were sent to his uncle Thomas in Dublin, is a major find for the topographical history of London. Though the series of letters as it now exists is not complete, it provides an intimate picture of some parts of the city, an intimacy based on the personal interests of their author and his connections through his family to some of the city's more interesting residents. But London is not the only subject of these letters: Molyneux also describes Oxford, with other places en route, such as Hampton Court and Windsor Castle, and some within the close environs of Oxford, particularly Blenheim Palace. (Molyneux's description of Cambridge is not so good, suffering as it does from the incompleteness of the manuscript.) Paul Holden, in his introduction to the Oxford letter, places Molyneux's descriptions in context by quoting other contemporary writings, especially those of Thomas Hearne, who had a less than flattering reaction to Molyneux. Oxford also provided Molyneux with the chance to see the Bodleian Library and to indulge his bibliophilia, a 'vice' which informs a good part of his recording of London sites.

Paul Holden is to be congratulated on this fine edition, as is the London Topographical Society on its publication. All historians of London who are interested in the early eighteenth-century city will surely echo these sentiments.

CLYVE JONES
Institute of Historical Research
University of London

SELECTED CONTENTS OF COPY-BOOK

The following list is of the more important sites visited by Molyneux when in London, Oxford and Cambridge. Lesser places are included in the full index.
Note: numbers appearing in bold throughout the introduction and commentaries relate to the folio numbers within the original manuscript.

ACKNOWLEDGEMENTS

This book was effectively born by chance. Whilst conducting research for a forthcoming work on Cornish country houses, I visited Southampton City Archives with the intention of examining a 1715 'Survey of the Duchy of Cornwall' by Samuel Molyneux, Secretary to the Prince of Wales. Realising that the document was of only limited use for my purpose, I took the opportunity to explore the Molyneux collections further, when I came across this manuscript.

On reading the opening folios I was immediately transported back into early Enlightenment England. As an architectural historian I was particularly drawn to the description of St Paul's Cathedral, but when the name 'Mr.Wren' (f. **36**) was indicated my pulse quickened. To my delight, as I read further the content got better and better; royal residences, public buildings, private institutions, splendid gardens and notable collections were all mentioned, some in great detail.

Realising the importance of this material, I suggested to the duty archivist that a transcription of the manuscript was needed. I owe a great debt of thanks to Miriam Phillips who painstakingly transcribed the whole manuscript and to Sue Hill and her team at Southampton City Archives for facilitating the transcription, providing illustrations for this volume and supporting its publication. Once the transcription was complete, I forwarded relevant sections to curators of the buildings and collections mentioned or authorities in their respective fields. I could not have been more surprised at their reactions which, without exception, commented that the source was unknown to them. I have incurred a debt of gratitude to the individuals who have helped me shape and contextualise Molyneux's visit to England. Specific thanks must be given to Bruce Barker-Benfield, Jonathan Betts, Alex Buck, Bridget Clifford, Joanna Corden, John Forster, Jonathan Foyle, John Goodall, Julian Harrison, David Hayton, Alastair Laing, Arthur MacGregor, Alison Morrison-Low, Lyndsy Nairn, Dan Pemberton and David Souden. I would like to thank Toby Barnard and Paul Cox for help in locating an image of Samuel Molyneux, Sheila O'Connell for help sourcing illustrations and contributing an epilogue and Kate Dinn for reading, correcting and commenting on my first draft. Staff of the John Rylands Library and university libraries in Exeter, Plymouth and Southampton have, as always, made research a pleasure. My employer, the National Trust, has been supportive of this work and provided images for which I am grateful.

I would like to single out two people whose help in taking this germ of an idea and shaping it into a book has been invaluable. First, to Clyve Jones, honorary fellow of the Institute of Historical Research and editor of *Parliamentary History*, who was instrumental in championing the significance of the manuscript. He commissioned articles based on Molyneux's visit to the Palace of Westminster[1] and the Harleian Library,[2] provided me with useful contacts and notified the London Topographical Society of the manuscript's existence. I am particularly thankful for Clyve's perceptive criticism, forbearance and encouragement and can think of no person more qualified for writing the foreword of this book. Secondly, I owe much to the observations of Ann Saunders of the London Topographical Society who has guided this work through publication and has written the preface. From the start Ann had unshakable confidence in both the source material and my abilities to interpret it. As an author it is so reassuring to have a sympathetic and supportive editor.

The two-year quest for Samuel Molyneux has proved exciting, thought-provoking and, at times, maddening. Therefore, I owe a debt greater than I can ever repay to my wife Kathryn and daughter Eleanor who have been long-suffering observers of this work — to them this book is dedicated.

PAUL HOLDEN
House and Collections Manager
The National Trust, Lanhydrock, Cornwall

Paul Holden has published widely on architectural history and curatorial issues in *Apollo*, *Country Life*, *Furniture History*, *Georgian Group Journal*, *Hampshire Studies*, *Journal of Liberal History*, *Journal of the Royal Institution of Cornwall* and *Parliamentary History*. His most recent publications include *The Lanhydrock Atlas* (Cornwall Editions, 2010) and '"Of Things Old and New": The Work of Richard Coad and James M MacLaren', in Jason Edwards and Imogen Hart (eds.), *The Aesthetic Interior* (Ashgate, 2010) and is currently working on a book on Cornish country houses. He was elected Fellow of the Society of Antiquaries in 2011.

1. Paul Holden, 'Westminster in 1712: A Description by Samuel Molyneux', *Parliamentary History*, 29, no. 3 (2010), pp. 452–9.
2. Paul Holden, '"One of the most remarkable things in London": A Visit to the Lord Treasurer's Library in 1713 by Samuel Molyneux', *British Library Electronic Journal* (2010) http://www.bl.uk/eblj/2010articles/article10.html.

INTRODUCTION

Samuel Molyneux (1689–1728) (Fig. 1) was a well-respected professional and intellectual who, between 1714 and 1727, dutifully served as principal secretary to the Prince of Wales, later King George II.[3] During his short career he represented constituencies in both the English and Irish parliaments, became an Irish Privy Councillor and, at the accession of George II in 1727, was created a Lord of the Admiralty.[4] In 1717 he married Elizabeth Capel (d. 1758/9), daughter of Lieutenant-General Algernon Capel (1670–1710), the 2nd Earl of Essex, and Lady Mary Bentinck. The marriage brought Molyneux great wealth, indeed his wife's dowry amounted to £10,000 and £18,000 followed in 1721 when Lady Elizabeth inherited Capel House (later the White House) at Kew in Surrey, from her aunt Dorothy, Lady Capel (*c.* 1642–1721), widow of Sir Henry, Lord Capel of Tewkesbury (1638–98).[5] Described in 1678 by John Evelyn as an 'old timber house' with a garden containing 'the Choicest fruit of any Plantation in England', the point was confirmed in 1724 by John Mackay when he added 'Mr Molyneux [is] said to have the best fruit in England collected by that great statesman and gardener my Lord Capel'.[6] In 1731 the contents of Capel House were sold for £720 8*s.* 6*d.*, the property was leased to James Pelham, Secretary to Fredrick, Prince of Wales, and between 1731 and 1735 it was rebuilt by the noted architect William Kent. After many years as a royal residence the house was demolished in 1802.

Despite having had a successful career at court, history has remembered Molyneux for two distinct reasons. At his home in Kew he built an observatory and rose to public prominence through his pioneering astronomical research in collaboration with Oxford University's Savilian

3. William A. Shaw and F. H. Slingsby (eds.), *Calendar of Treasury Books*, 29 (London, 1957), pp. 247–8 and 335–50. On 11 January 1715, £5,000 was approved by the House of Lords for Molyneux's use in the service of the Prince of Wales. In John Chamberlayne, *Magnae Britanniae notitia, or the Present State of Great Britain* (London, 1723), p. 562, Molyneux is recorded as Principal Secretary to His Royal Highness's family, taking a Patent fee of £66 6*s.* 8*d.* and a salary of £640 per annum.
4. Molyneux served as Member of Parliament for Bossiney and St Mawes in Cornwall and Exeter in Devon shortly before his death. For a political career, refer to Romney Sedgwick, *The House of Commons 1715–1754, Members E–Y* (London, 1970), p. 263.
5. Sir Capel Molyneux, *An account of the family and descendants of Sir Thomas Molyneux* (Dublin, 1820), p. 38. A rough equivalent today (2011), using the National Archives currency converter, would be £765,000 and £1,525,500 respectively, http://www.nationalarchives.gov.uk/currency/results.asp#mid [accessed 5 December 2010].
6. E. S. de Beer (ed.), *The Diary of John Evelyn* (London, 2007), p. 583.

Fig. 1. Samuel Molyneux, artist unknown, 1727. Inscribed upper left 'Rt. Hon. Mr Secretary / Molyneux', repeated lower right with the addition 'Member for the University / 1727'. From the collection of Mrs Molyneux of Trewyn (1958), later sold at auction, whereabouts now unknown.

Photograph © National Portrait Gallery, London; unknown collection

Professor of Astronomy, James Bradley (1693–1762). Jointly they designed and produced optical instruments with which Bradley later discovered the effect of the stellar aberration.[7] Yet his reputation as a man of science was severely injured in 1726 when he applied his sceptical and analytical mind to the claim that Mary Toft from Godalming was giving birth to live rabbits.[8] Molyneux, along with the Swiss surgeon and anatomist Nathaniel St André (1680–1776), was deceived into believing the claim was true, and when it was eventually declared as fraudulent a public scandal ensued with both men being venomously ridiculed. The height of public derision came first with William Hogarth's print entitled *Cunicularii*, published on

7. A. D. C. Simpson, 'The Beginnings of Commercial Manufacture of the Reflecting Telescope in London', *Journal of the History of Astronomy*, XL (2009), p. 433. Thanks to Alison Morrison-Low for the reference.
8. Mr St André, *A Short Narrative of an extraordinary Delivery of Rabbets perform'd by Mr John Howard* (London, 1726). Mary Toft confessed the fraud on 7 and 8 December 1726 and St André published a retraction in the *Daily Journal* on 9 December 1726. Molyneux was also derided by Jonathan Swift (publishing under the pseudonym of Lemuel Gulliver) in a pamphlet entitled *The Anatomist dissected; or the Man-Midwife finely brought to Bed. Being an examination of the conduct of Mr St André touching the late pretended Rabbit-bearer; as it appears from his own Narrative* (London, 1727). For modern interpretations, see Dennis Todd, *Imagining Monsters: Miscreations of the Self in Eighteenth-Century England* (Chicago, 1995) and Clifford A. Pickover, *The Girl Who Gave Birth to Rabbits* (New York, 2000).

26 December 1726, and secondly through a poem penned by Alexander Pope entitled *The Discovery: or, The Squire turn'd Ferret* (see Epilogue).

Consequently, history has critically misjudged Samuel Molyneux as a discredited courtier and ineffectual politician, overshadowing his formative years as a talented young scholar of antiquities, history, nature, philosophy and science. This book aims to redress this imbalance by presenting for the first time a complete transcription of seven remarkable letters written between December 1712 and April 1713 when Molyneux was in England in pursuit of edification and self-improvement beyond his native Ireland. These letters provide first-hand accounts of some of the greatest connoisseurs of the age, their collections and some of the best buildings of the early Enlightenment period in London, Oxford and Cambridge. They portray the author without disguise, as a natural philosopher and scientist, a young man engrossed in the past whilst showing belief and hope for the future, and an antiquary with one eye on collecting and classification and the other on the conjectural history of the free-thinking world. Such enlightened thought not only liberated inhibitions in challenging certain established hypotheses but also raised expectations to meet like-minded, well-educated, wealthy connoisseurs who had followed their own obsessive passions for collecting works of art and antiquities. Like John Loveday's diaries, written between 1729 and 1765, Molyneux's correspondence imparts a learned and far-reaching dialogue based on the intellectual elitism of the period, quite unlike other contemporary tourists such as Zacharias Conrad von Uffenbach (1710), Daniel Defoe (1724) and Don Manoel Gonzales (1731), who for the most part commented on London's domestic customs and its political, social, theological and topographical arrangements.[9]

Samuel Molyneux (1689–1728)[10]

Samuel Molyneux was the only surviving child of the astronomer, antiquarian, philosopher and constitutional writer William Molyneux

9. W. H. Quarrell and Margaret Mare (eds.), *London in 1710: From the Travels of Zacharias Conrad von Uffenbach* (London, 1936); P. N. Furbank, W. R. Owens and A. J. Coulson (eds.), *Daniel Defoe: A Tour Through the Whole Island of Great Britain* (New Haven and London, 1991); Don Manoel Gonzales, *London in 1731: Containing a description of the city of London; both in regard of its extent, buildings, government, trade &c* (London, 1888); Sarah Markham, *John Loveday of Caversham 1711–1789; The Life and Tours of an Eighteenth Century Onlooker* (Salisbury, 1984).
10. John Bergin, 'Samuel Molyneux', in James McGuire and James Quinn (eds.), *Dictionary of Irish Biography Online* (Cambridge: Royal Irish Academy, 2010), www.dib.cambridge.org [accessed 23 June 2010]; A. M. Clerke, revised by Anita McConnell, 'Samuel Molyneux (1689–1728)', in *Dictionary of National Biography Online* (Oxford, 2004), www.oxforddnb.com/view/article/18925?docPos=13 [accessed 5 May 2009]. For the scientific achievements of William and Samuel Molyneux, see Charles Coulston Gillispie (ed.), *Dictionary of Scientific Biography*, IX (New York, 1974), pp. 463–6.

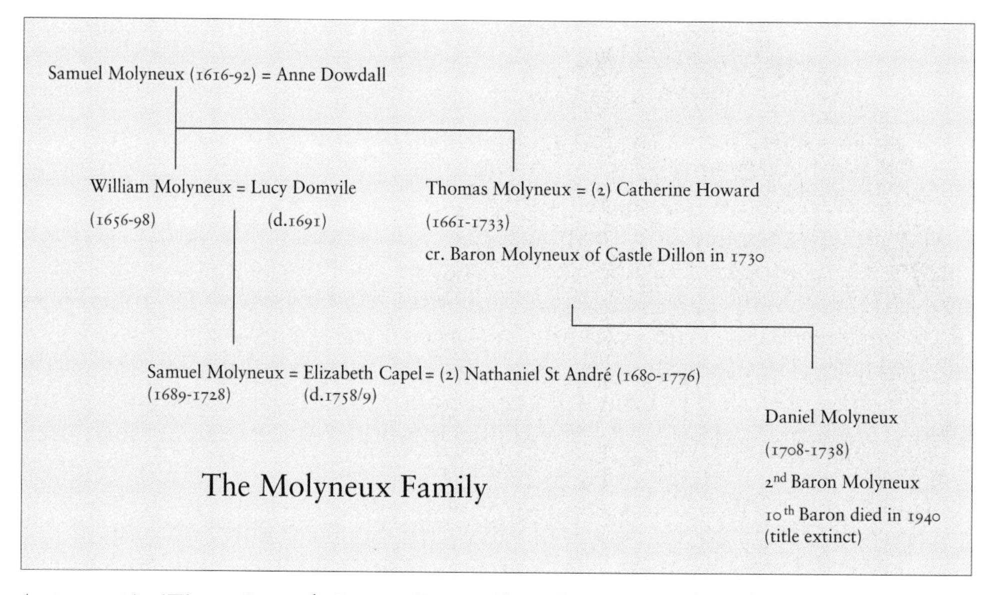

The Molyneux Family

(1656–98) (Fig. 2) and Lucy Domvile (d. 1691), daughter of the Irish Attorney-General. According to Jonathan Swift, William Molyneux was 'an English gentleman born in Ireland [who] never grew tired of proclaiming the fact' — indeed, his great-grandfather, Thomas, was English and was Queen Elizabeth's nominee for the Chancellorship of the Irish Exchequer in 1576.[11] The family's interest in antiquarianism may well have stemmed from William's grandfather, Daniel (d. 1632) who was appointed Ulster King-of-Arms in 1586, while its wealth derived from his father, Samuel Molyneux (d. 1692) who, being an expert in ordnance, rose to the senior position of Ireland's Chief Engineer.

William Molyneux was educated at Trinity College, Dublin, where he received his Bachelor of Arts degree before studying at Middle Temple between 1675 and 1678 (Fig. 2). His scientific and published research attracted friendships with Royal Society notables such as Robert Boyle, John Flamsteed, Edmund Halley, Robert Hooke, Christiaan Huygens, John Locke and Isaac Newton.[12] In 1683, from a coffee-house on Cork Hill, he founded with Sir William Petty the 'Dublin Philosophical Society for the Improvement of Natural Knowledge, Mathematics and Mechanics'. On 3 February 1686, after he had travelled to the Low Countries with his brother, Thomas Molyneux, he was elected as a Fellow of the Royal Society.[13] With

11. Quoted in Joep Leerson, *Mere Irish and Fior Ghael* (Amsterdam, 1986), p. 350.
12. Correspondence between John Flamsteed, Astronomer Royal of the Greenwich Observatory, and William Molyneux of Dublin is held in the Southampton City Archives (hereafter SCA) D/M 1/1.
13. 'Gallery of Illustrious Irishmen, No.XII, Sir Thomas Molyneux, Bart, part 1', *Dublin University Magazine*, 18 (1841), pp. 305–27.

Fig. 2. William Molyneux attributed to Sir Godfrey Kneller, Bt., *c.* 1696.

© National Portrait Gallery, London

the imminent threat of Catholic nationalist resurgence and consequent persecution from the Lord Deputy Richard Talbot, the Earl of Tyrconnell, William and his wife fled from Ireland to Chester.

Victory at the Battle of the Boyne for King William III and the return of the Protestant ascendancy prompted William, with the newborn Samuel in tow, to return to Dublin. After his wife's death in 1691 he sought solace in his scientific and constitutional research, the culmination of his efforts being his 1698 published tract *The Case of Ireland being bound by Acts of Parliament, Stated.* This influential tract asserted Ireland's position as an independent nation 'governed', he stated, 'only by such laws to which they give their own consent by their representatives in parliament'. A parliamentary union with England, he added, would have been 'a happiness we can hardly hope for'.[14] The work was considered so inflammatory by the English parliament that it ordered its destruction, yet the tract proved so significant that it was reprinted nine times between 1706 and 1782. Within the year of its first publication William Molyneux was dead.

The nine-year-old Samuel was brought-up in Dublin by his uncle Thomas Molyneux (1661–1733), who had received an MA from Trinity College, studied medicine in Holland, travelled abroad and was in England between

14. William Molyneux, *The Case of Ireland being bound by Acts of Parliament, Stated* (Dublin, 1698), pp. 48 and 97–8.

1683 and 1687 when he was elected a Fellow of the Royal Society in 1686, nine months after his brother, William. After he returned to Dublin from Chester he forged a successful career as a physician, being elected a Fellow of the Irish College of Physicians, becoming President in 1713, and later serving as Regius Professor of Physic in 1717. At his Dublin home Thomas housed a noted collection of paintings, a significant collection of natural history specimens and a well-stocked library from which he wrote many published works in fields of archaeology, antiquarianism and natural history. For the young Samuel, his late father's reputation along with his uncle's discernment in antiquarianism must have been a great motivation for his own future.

As was the family tradition, Samuel studied at Trinity College where he became great friends with the mathematician and philosopher George Berkeley (1685–1753). He matriculated in 1705, took a BA in 1707, an MA three years later and in November was elected secretary to the newly-revived Dublin Philosophical Society. The Bishop of Clogher wrote with enthusiasm:

> Sir I doe very heartily congratulate you on your being chosen Secretary to the [Dublin] Philosophical Society. Your worthy Father was our first Secretary upon our Establishment, and you are so upon our Revival. May you still succeed to and inherit every one of your father's valuable good qualitys.[15]

In pursuit of antiquarianism, astronomy, history, natural philosophy and empirical science, Samuel travelled throughout Ireland, published his research within the Society's Journal and benefited from the social and intellectual freedom of membership.[16] To advance his studies Molyneux used his late father's and uncle's exalted reputations to further his own ends. On 24 June 1707, in a letter to the distinguished scientist, Francis Hauksbee, he actively sought an introduction to 'the ingenious Mr Derham'[17] who was 'well known to the Gentlemen of the Royal Society', writing:

15. SCA D/M 1/2, letter dated 30 November 1707.
16. Samuel Molyneux, 'A Relation of the Strange Effects of Thunder and Lighting, Which Happened at Mrs Close's House at New-Forge, in the county of Down in Ireland, on the 9th August 1707. Communicated by Samuel Molyneux Esq, Secretary of the Philosophical Society at Dublin', *Dublin Philosophical Society Transactions* (hereafter *DPST*), 26 (1708), pp. 36–40. Lord Archbishop of Dublin and Samuel Molyneux, 'An Account of the Manner of Manuring Lands by Sea-Shells, as Practised in the Counties of Londonderry and Donegall in Ireland, By His Grace the Lord Archbishop of Dublin. Communicated by Samuel Molyneux Esq', *DPST*, 26 (1708), pp. 59–64. His Irish travels and research papers are held in the archives of Trinity College, Dublin (hereafter TCD). Samuel Molyneux, 'Journey to the North', TCD MS 888/2, ff. 183–8, reproduced in R. M. Young, *Historical Notes of Old Belfast* (Belfast, 1896), p. 152; Samuel Molyneux, 'Natural History of Ireland &c', TCD MS 883; Samuel Molyneux, 'Journey to Connaught', April 1709, TCD 884, reproduced in A. Smith (ed.), *Miscellany of the Irish Archaeological Society*, 1 (1846), pp. 161–78.
17. William Derham (1657–1735), naturalist, physician and Fellow of the Royal Society.

> Sir I know not whether my Father, Wm. Molyneux, had ye honour of being known to you during his life. If he had the happiness of your acquaintance, I beg it may in some way excuse the ill manners of his son in thus troubling with an impertinent letter a person no otherwise known to him than as he is to the whole learned world by his great worth and learning.[18]

In December 1707 he corresponded with 'the Ingenious Dr. Hans Sloan, Secretary to the Royal Society', in an effort to rebuild links between the Royal Society in London and the Dublin Philosophical Society, yet despite Molyneux's best efforts the Society fell apart after his departure for England.[19]

In the light of surviving evidence we can build a fair picture of Molyneux's character and temperament. His own father thought him a 'precocious talent', Lord Egmont remarked on his loyalty, Lady Cowper regarded him as untrustworthy and the family historian, Nellie Zada Rice Molyneux, later summed him up as '[...] a man of winning manners and obliging temper [... who] united Irish wit to social accomplishments [while his] inflexible integrity seemed to stand in his way to social advancement'.[20] 'Winning manners' through civility and virtue were promoted strongly throughout Molyneux's Lockean education and expected in polite Whiggish society. Indeed, *The Spectator* noted, 'It ought certainly to be the first point to be aimed at in Society, to gain the good Will of those with whom you converse'.[21]

Despite his polite demeanour, at several points in the letters his 'inflexible integrity' is laid bare through his obstinate devotion towards the Protestant faith (approximately 55 per cent of the 70,000 Dublin residents in 1710 were Protestant) and his support for the Whig party. Molyneux wholeheartedly shared his father's political views, in particular that the Glorious Revolution and the Protestant Succession were milestones towards ending the threat of Jacobite resurgence, by bringing about providential delivery from popish tyranny. In his diary entry for 3 March 1713, Thomas Hearne, who was Molyneux's guide to the Bodleian Library during the latter's two-day trip to Oxford, leaves no doubt as to the Irishman's political allegiance, obstinate temperament and unyielding self-confidence. The peevish Bodleian under-librarian described him as 'a Man of some Confidence' and a 'very Whiggish Gentleman', a point confirmed by the company he kept that day,

18. SCA D/M 1/2.
19. For further correspondence with Hans Sloane, refer to British Library (hereafter BL), Sloane MS BM 4041, f. 190, 4044, f. 274 and 4036–41.
20. Nellie Zada Rice Molyneux, *History of the Molyneux Families* (New York, 1904), p. 145.
21. For more on Molyneux's education, refer to John Locke, *Some Thoughts Concerning Education* (London, 1693) and *The Works of John Locke*, IX (London, 1823), pp. 289–472. *The Spectator*, No. 422, 4 July 1712, p. 103.

namely Mr Keill, the Professor of Astronomy, regarded by many as one of the most vehement and violent Whigs in Ireland.[22] Molyneux therefore fits well with Swift's description of a typical Whig, as he

> [...] profess'd himself to approve the *Revolution*, to be against the *Pretender*, to justify the Succession in the House of *Hannover*, to think the *British Monarchy* not absolute, but limited by Laws, which the Executive Power could not dispense with, and to allow an Indulgence to Scrupulous Consciences; such a man was content to be called a *Whig*.[23]

When Hearne challenged Molyneux's uncompromising loyalty to the Crown by showing him a picture of James Francis Edward Stuart, the Prince of Wales (Queen Anne's half-brother, known as the 'Old Pretender'), so incensed was Molyneux that he complained to Dr Charlett who consequently reprimanded Hearne.[24]

This unpleasant incident brought retaliation in the form of a venomous attack on the Irishman's scholarly acumen. His 'Ignorance in Antiquities', Hearne wrote, was based on his 'superficial' knowledge taken from 'Accounts from Conversation with Gentlemen, and not from study'. He added:

> [...] tho' after all, I was willing to think as favourably of him, in this respect, as I could, at least to believe, that this was not so much the Effect of Conceit, as of some considerable Insight into Antiquities; which might make him talk a little more lavishly of Affairs of this kind than Persons of less Improvement in these Studies. I therefore submitted to what he asserted, yet with this Reserve, that I should change my Opinion of his Abilities whenever I found that he had not cultivated these Studies with that Industry and Application as one would by his Discourse gather that he had done. [...] Hitherto I had sufficient and full Proof of Mr. Mollineux's Confidence and of his Ignorance in Antiquities. Yet he had not as yet discovered himself to be a Man of Republican, ill Principles, and of a malignant Temper [...][25]

An element of Hearne's analysis rings true when the whole manuscript is considered. Two further examples illuminate Molyneux as a theoretical enthusiast rather than a practical authority. First, his account of some of the books within the Harleian Library (ff. **98–102**) follows closely the preface entries of the library catalogues published between 1808 and 1812

22. D. W. Rannie (ed.), *Remarks and Collections of Thomas Hearne, vol. IV*, XXXIV (Oxford Historical Society, 1898), p. 92.
23. *The Examiner* (Editorial), 24–31 May 1711.
24. Rannie, *Hearne*, p. 165. Hearne initially reported that Molyneux had complained to Dr Gardener, the Warden of All Souls, but changed this in his diary entry for 24 April when he wrote, 'Mr Keil assured me yesterday that Mr Mollineux complained to Dr Charlett'.
25. Ibid., pp. 110–12.

of which his host, Harley's 'Library-keeper' Humfrey Wanley (1672–1726), had compiled to number 2,407.[26] Second, while in the company of Thomas Hearne in Oxford, Molyneux had cause to recall having seen a genuine Brass Otho (a rare Roman imperial coin) in the Earl of Pembroke's study. Hearne wrote disbelievingly, 'This made me increase my Opinion of his Confidence; however I said no more than this, that, with Submission, I much questioned it'.[27]

In consideration of Molyneux's civility, poise, loyalty to the Hanoverian succession and Whiggish politics, it can be of little surprise that he steadfastly served the Crown from 1714 nor, taking into account his pedigree, that in 1715 he was listed alongside 'Meath, Mountjoy, Castlecomer, Robt Finlay, [and] Richard Steele' as 'stewards of the Protestants of Ireland for the celebration of their anniversary meeting in London, in memory of their deliverance from a general massacre begun in that kingdom by the Irish Papists in the year 1641'.[28] Perhaps it was these very qualities that so irritated his fellow Irishman Jonathan Swift (1667–1724) when he recorded in a 'Letter to Stella' that 'I presented Pratt (Provost of Trinity College, Dublin) to Lord-Treasurer, and truly young Molyneux would have had me present him too; but I directly answered him I would not, unless he had business with him'.[29]

Molyneux in London

'Miserable and divided' was how Daniel Defoe described the nation that Samuel Molyneux entered in October 1712, probably for the first time since his childhood.[30] Constitutionally and politically, the prospects of the nation were delicately poised and highly charged, a consequence of party rivalries that split the nation and created instability. A Tory Government had replaced the Whigs in October 1710, yet despite having a 200-seat majority in the House of Commons the Lords was still dominated by the Whigs. To make matters worse, the Tories had split into several factions and consequently remained in total disarray over the question of the Hanoverian Succession and the War of the Spanish Succession, of which the latter caused a great strain on the public finances. The Whigs meanwhile defended the war, which had been fought almost continually from 1689, and stood firm behind the 1701 Act of Settlement that decreed in the event of King William III or his

26. *A Catalogue of the Harleian Manuscripts in the British Museum*, 4 vols (London, 1808–12).
27. Rannie, *Hearne*, p. 109.
28. *Manuscripts of the Marquess Townsend*, Historic Manuscripts Commission (London, 1887), p. 130.
29. Temple Scott (ed.), *The Prose Works of Jonathan Swift (Vol. II: The Journal to Stella 1710–1713)* (London, 1922), p. 390. In the same letter Swift expressed his low opinion of William Molyneux whom he considered 'a dangerous author'.
30. Daniel Defoe, *Review*, VIII, p. 489.

heir Anne defaulting on legitimate issue, the Protestant Sophia, Dowager Electress of Hanover and granddaughter of King James I of England, would succeed to the English throne. Any such settlement would effectively end any claim on the English throne by the Catholic James Francis Edward Stuart, despite his legitimate claim on the English throne by reason of primogeniture. The Tory government was returned in September 1712 with only a slight reduction in its majority and, to the chagrin of many Whigs, established peace through the ratification of the Treaty of Utrecht in March 1713, which in the eyes of many betrayed the House of Hanover and its allies.

Religious opinion and the treatment of religious minorities, too, aroused deep division. Molyneux, who was educated under Lockean philosophical theories of natural rights and religious toleration, does not discuss theological matters in his letters, but does refer to meetings with 'the Archb^{p} of Canterbury, the Bishops of Ely, Bangor & Litchfield' (f. **148**). These meetings may reflect more the expertise and discernment in antiquarianism of the participants than any theological interest on Molyneux's part. His curiosity was, however, aroused enough to attend prayers said by the 'Fam'd D$^{r:}$ Sacheverell' (f. **148**), the High Church divine who in 1709 preached two scandalous sermons that indited a corrupt Church and Government.[31] Sacheverell's subsequent trial evoked strong opinions which often erupted into violence, and he was eventually banned from sermonising for three years. Disapproving of Sacheverell's principles, Molyneux wrote, 'if his Doctrine were agreeable to his voice they would be full of the softest and most effectual Perswasion' (f. **148**).

Culturally, however, things were far from miserable. The interest in antiquarianism led to a trade in antiquities, creating a new generation of collectors and connoisseurs. Along with the commercialisation of the printing press, a new confidence in the arts was established that embedded itself within eighteenth-century society. Opulence gave way to elegance, magnificence to refinement, and serious, well-educated and much-travelled collectors defined fashions in taste. Such intellectualism was not exclusive to the landed elite but bridged social, political and religious divides, spawning a movement led by academics, antiquarians, artists, bibliophiles, collectors, historians, scientists and writers, later known as the Enlightenment. *The Spectator* extended the mandate when, between 21 June and 3 July 1712, it published eleven thematically-linked essays entitled 'The Pleasures of the Imagination' that, for the middle classes, reinforced the case for a moral, philosophical and ideological approach towards art,

31. The first, *The Peril of False Bretheren*, was delivered before the Lord Mayor in St Paul's Cathedral on 5 November, and the second, *Communication of Sin*, was delivered to the Derby Assizes on 13 December.

architecture, history, literature and nature.[32] Inspired by this development in the arts, Molyneux decided to visit London, Oxford and Cambridge — three places where enlightened thought could deepen and where libraries and learned societies could flourish.

One such society symbolic of the transformation of English intellectual life was the Royal Society, in which Molyneux was enrolled as a Fellow on 1 December 1712. Amongst its ranks were virtuosi and polymaths with an insatiable appetite to challenge established scientific theories through the application of logic and experimentation. Curiously, this key event in Molyneux's life is not mentioned in his letters, although it is recorded in the Royal Society's Journal books and council minutes, as is his attendance, 'at his request', to a scientific meeting at the Royal Society on 30 October 1712.[33] On 4 November Molyneux, along with 'Mr Bernoulli [...] and Mr Turner', were proposed for election as Fellows of the Society; 'Mr Bernoulli and Mr Molyneux were approved of, and Mr Turner was disapproved'. On 6 November Molyneux was again at the Society. This time he addressed the meeting, giving descriptions of cross-bills, as seen in the north of Ireland, and of Irish marble, noting that 'the best marble came from the County of Kerry, that the white marble turns yellow, and that the Irish marble was brittle and full of cracks'. The Journal book entry for 1 December 1712 records 'This being Election Day of the officers of the Society, the 30th of November happening on a Sunday. Mr Molyneux, Dr Bernoulli, Mr Tempest, Dr Patrick Blair and Mr Bradley were Proposed, Ballotted for, and Chosen Members of the Society'.[34] In 1701, Ralph Thoresby gave some idea of the formality of the process, describing '[...] the Vice-President's taking me by the hand and publicly pronouncing me (in the name of the Society) a Fellow of the Royal Society'.[35] Of the processes involved, Don Manoel Gonzales wrote in 1731

> When a gentleman desires to be admitted to this society, he procures one of the Corporation to recommend him as a person duly qualified, whereupon his name is entered in a book, and proper inquiries made concerning his merit and abilities; and if the gentleman is approved of, he appears in some following assembly, and subscribes a paper, wherein he promises that he will endeavour to promote the welfare of the society, and the president formally admits him by saying 'I do, by the authority and in the name of the Royal Society of London for improving natural

32. *The Spectator*, No. 411, 21 June 1712, to No. 412, 3 July 1712, pp. 62–103.
33. Thanks to Joanna Corden of the Royal Society for transcribing the Royal Society Journals and minute books.
34. Jean Bernoulli (1667–1748) Swiss mathematician, William Tempest (1682–1761) barrister, Patrick Blair (*fl.* 1708–28) physician and botanist and Richard Bradley (d. 1732) botanist.
35. Joseph Hunter (ed.), *The Diary of Ralph Thoresby* (London, 1830), p. 339.

> knowledge, admit you a member thereof'. Whereupon the new member pays forty shillings to the treasurer, and two-and-fifty shillings per annum afterwards by quarterly payments, towards the charges of the experiments, the salaries of the officers of the house &c.[36]

Molyneux's enrolment into the Society gives a clear reason for his being in London, but it appears not to have been his sole purpose. By 20 April 1713 the letters place Molyneux at Harwich (f. **156**), where he 'thence [sailed] directly to Holland' (f. **154**) in the service of the Duke and Duchess of Marlborough. In an age when espionage was prevalent, it is conceivable that he was employed as a London agent (or spy) before finding a commission abroad — indeed, it may be argued that the factual and systematic nature of the letters may potentially have been some kind of ruse to outmanoeuvre suspicious authorities. His pedigree, Royal Society fellowship, 'winning manners' and good reputation as a learned antiquarian would have proved very plausible, so much so that it resonates with Daniel Defoe's enthusiasm for his own spying skills when he wrote to Lord Oxford that 'I converse with Presbyterian, Episcopal-Dissenter, papist and Non-Juror, and I hope with equall circumspection. I flatter myself you will have no complaints of my conduct. I have faithful emissaries in every company and I talk to everybody in their own way'.[37]

How Molyneux first became involved with the Marlboroughs remains unclear. In terms of recommendation, as secretary of the Dublin Philosophical Society, Molyneux would have been familiar with its influential President Lord Pembroke, Lord Lieutenant of Ireland in 1707, who he describes in his letters as 'a good friend' (f. **143**), and with Andrew Fountaine, one of Pembroke's officials at Dublin Castle who provided Molyneux with at least one introductory letter. Other mutual contacts included Frances, Lady Tyrconnell, sister of the Duchess of Marlborough; Joseph Addison, chief secretary to Lord Wharton Lord Lieutenant of Ireland in 1709; Archbishop King, who advised Molyneux to study Italian and French four months before going abroad;[38] and the Irishman William Cadogan (1672–1726), a fellow Trinity College student and one of Marlborough's closest aides.

Since arriving in London, Molyneux recorded that 'I have had the Honour since I have been here to be sometimes receiv'd by the Duke of Marlborough [...]' (f. **148**), but where and when these meetings took place remains vague. The 1820 published family history states that he first visited Marlborough at Antwerp in 1712, which is improbable as the duke arrived in Ostend on

36. Gonzales, *London in 1731*, p. 42.
37. George Harris Healey (ed.), *The Letters of Daniel Defoe* (Oxford, 1955), pp. 158–9.
38. Archbishop King to Samuel Molyneux, TCD 750/4/1/73–4 (25 November 1712) and TCD 750/4/1/100–2 (6 January 1713).

1 December 1712 (to court the heir presumptive to the British throne), when Molyneux was clearly still in London.[39] October 1712 presents itself as the most likely opportunity when the pair first met, either at the duke's Holywell estate near St Albans, which Molyneux could have passed en route to London (ff. **30–2**), or when the duke was in London acting as a pallbearer at the funeral of his great friend Lord Godolphin, who had died at Holywell on 15 September and was buried on 7 October 1712. In keeping with the embargo he seems to have imposed on political and social news in his letters, Molyneux mentions neither the death of Sidney, the 1st Earl of Godolphin, nor the allegations in the press that Marlborough somehow encouraged the scandalous duel between the Duke of Hamilton and Lord Mohun which resulted in their deaths on 15 November 1712, Hamilton by Mohun and Mohun by Hamilton's second.

The letters show no direct evidence of aristocratic support or patronage whilst Molyneux was in England. Political infighting forced Marlborough's exile abroad on 24 November 1712, a point Archbishop King noted in a letter to Molyneux on 25 November when he wrote that 'the great man [is now] out of Great Britain'.[40] Interestingly, Swift commented on the prospect in a 'Letter to Stella' as early as 28 October 1712, which may suggest that Molyneux had mentioned the fact when they met, most likely at the meeting of the Irish Protestants in London.[41] If not retained by the duke at this stage, Molyneux was certainly a supporter as shown in an undated letter from Molyneux to his cousin which states:

> I have passed the winter most agreeably here, at Antwerp, whither you know the Duke and Duchess of Marlborough have retired; one would think, by the many civilities I have received from them, that I had deserved as much from them as they have from the public; but as this is impossible, I must content myself only in endeavouring to deserve them, and in taking a pleasure to assure all my friends of my gratitude and obligations to these two great persons.[42]

Rarely in the manuscript does he express any partisan views, but, after seeing the military standards on display in Westminster Hall, he reflects on the duke's exile, writing on 20 December, 'Who will believe that Gratitude can be so much a Stranger to Our Hearts that Spoyls like these can unregarded hang before a nation's face at home where he that won 'em for these very Spoyls now reigns a Prince but yet a banish'd man abroad' (f. **42**). Molyneux also shows concern for the immeasurable damage to the

39. Molyneux, *An account of the family and descendants of Sir Thomas Molyneux*, p. 37.
40. TCD 750/4/1/73–4.
41. Scott, *Swift*, p. 390.
42. Quoted in Molyneux, *An account of the family and descendants of Sir Thomas Molyneux*, p. 37.

duke's reputation through the unfinished building of Blenheim Palace, considering the disgraceful situation as 'little an Honour to the Nation as to the Builder that undertook it' (f. **139**).

Although patronage would have created opportunities and provided introductions to some of the foremost collectors of the day, Molyneux's own contacts cannot be underestimated. His pedigree, and his associations with the Dublin Philosophical Society and Royal Society clearly influenced his access to collections. Amongst those ranked as friends were fellow antiquarians Francesco Bianchini (f. **78**) and Lord Pembroke (f. **143**), and literary figures Christopher Bateman (f. **151**) and Joseph Addison (f. **152**). Some visits were arranged through letters of introduction or conducted either in company or with exclusive guides, while his membership of a learned society (f. **154**) and disapproval of boisterous coffee houses tells us much about his disposition.

Molyneux left London on 15 April 1713 (f. **156**). He was not to return until June 1714, when it was reported 'Young Molinex is come back from Hanover to Antwerp and is now going straight over to England by Ostende. Mr Harley knows his character'.[43] The Harley referred to here would appear to be Thomas Harley who was sent to Hanover by his cousin Robert, the Lord Treasurer. Although little documentation survives regarding Molyneux's time at the Hanoverian court, this closing comment refers to one revealing incident in which he was involved. Back in England the reputation of the Lord Treasurer, Robert Harley (1661–1724), had not weathered well. Concerned by the prospect of a Hanoverian succession, the Tory, who had loyally served Queen Anne as Speaker of the House of Commons (1701–5), Secretary of State (1704–8), Chancellor of the Exchequer (1710–11) and Lord Treasurer (1711–14) and was created the 1st Earl of Oxford and Earl Mortimer in 1711, seriously considered his loyalties and failed in his attempt to form a non-party administration. Jonathan Swift indicated Harley's position when he wrote to Archbishop King on 28 March 1713: 'The other Party are employed in spreading a Report most industriously, that the Lord Treasurer intends after the peace, to declare for the Whigs'.[44] Fearful for his future, Harley attempted to gain favour with the eighty-three-year-old Sophia, Dowager Electress of Hanover, by sending her a series of unsolicited letters.[45] Molyneux witnessed the Dowager's death in the Herrenhausen garden in June 1714, he wrote:

43. *Report on the Manuscripts of His Grace the Duke of Portland preserved at Welbeck Abbey, vol. IX*, Historic Manuscript Commission (London, 1923), p. 402.
44. David Woolley (ed.), *The Correspondence of Jonathan Swift, D.D.*, I (Frankfurt, 1999) p. 471.
45. Letters dated 1714 between Molyneux and the Duchess of Marlborough can be found at BL Add MSS (1714) 61465, ff. 5–21b and (1714) 61635, f. 131.

> And there the first thing I heard was that the good old electress was dying in one of the public walks. I ran up there, and just found her expiring in the arms of the poor electoral princess, and amist the tears of a great many of her servants, who endeavoured in vain to help her.[46]

He then executed her dying wish that the letters should be passed to the Duchess of Marlborough, who subsequently saw to it that they were published, effectively eroding Harley's support. As a result, it has been suggested that Molyneux was in fact employed by the duchess.[47]

Molyneux's Death

In March 1728 Samuel Molyneux, the newly elected Member of Parliament for Exeter, collapsed in the House of Commons. He was quickly attended to by Nathaniel St André, but died on 13 April 1728.[48] On 20 April 1728 the *Daily Journal* reported, 'On Thursday night last the corpse of the Hon Samuel Molyneux Esq was carried from his late Dwelling house in George Street in Hanover Square to be interred at Kew Green'.[49] As had been the case in the last years of his life, his death also aroused controversy; a published letter by the Revd Samuel Madden (1686–1765) insinuated that it was St André who had effectively administered a fatal dose of opium — a murderous charge that was later retracted.[50] Molyneux, himself a bibliophile and active collector, had inherited a fine library from his father, many volumes being brought across from Dublin in 1716 and temporarily stored in a hired room in Pall Mall.[51] After his death his library and manuscripts were bequeathed to his wife Elizabeth, who had eloped with St André and fell from favour with the queen as a result.[52] In Southampton, the pair married and developed an estate called Belle Vue on the outskirts of the city where they lived until their deaths, Elizabeth in 1758/9 and her husband in 1776. Molyneux's collections subsequently passed to St André's

46. William Coxe, *Memoirs of John Duke of Marlborough*, 3 (London, 1819), p. 574.
47. E. Gregg, 'Marlborough in Exile 1712–1714', *Historical Journal*, XV, no. 4 (1972), pp. 593–618, and Ophelia Field, *The Favourite: Sarah Duchess of Marlborough* (London, 2002), pp. 334 and 336.
48. Molyneux's father (and later his uncle and cousins) were buried in the family burial vault in the north aisle of St Audoen's church, High Street, Dublin.
49. *The Daily Journal*, 20 April 1728, issue 2269.
50. *Reverend Mr Madden to the Hon Lady Molyneux, on occasion of the death of the Rt Hon Samuel Molyneux Esq: who was attended by M St Andre a French Surgeon* (Dublin, 1730). Madden was married to Mary Molyneux, sister of William and Thomas Molyneux. For St André, see *Gentleman's Magazine*, July 1781, pp. 320–33.
51. SCA D/M 3/1 (7 January 1716); see also *A Catalogue of the Pitt Collection* (Southampton, 1964).
52. *A catalogue of the library of Samuel Molyneux deceas'd ... consisting of many valuable and rare books ... with several curious manuscripts* (London, 1730).

maidservant Mary Pitt, who bequeathed them to her children William and George, who eventually left the bulk of the collection by deed and gift to the Corporation of Southampton in 1831.[53]

The Manuscript

Amongst the Molyneux papers, now held in the Southampton City Archives, is a manuscript copy-book consisting of copies of seven letters from Samuel Molyneux to an unnamed recipient. The first of these (covering ff. **30–2**) is incomplete and undated, but was probably written in October 1712; the others are dated London, 20 December 1712 (ff. **33–43**, **58–63** and **77–82**), 14 February '1712/13' (ff. **83–97**), 18 February '1712/3' (ff. **98–104**), 28 February '1712/3' (ff. **105–12** and **118–46**), 15 April '1712/3' (ff. **147–54**) and Harwich, 20 April 1713 (ff. **156–9**). The dating of the letters follows the practice of the day whereby, as one contemporary wrote, '[the year] begins [...] from the day of Christ's incarnation on the 25th of March'.[54] The reason for copying the letters remains unclear.

The manuscript is incomplete, being foliated **30–43**, **58–63**, **77–112** and **118–59**. Theoretically, assuming approximately 260 words per page, some 7,800 words appear to be missing from the start of the manuscript and therefore by default after folio **159**. As his narrative starts with his journey to London and concludes in Cambridge en route to Harwich, it is likely that the manuscript only deals with his account of his time in England. The location of the original letters has not been traced.

The manuscript comprises folded paper sheets measuring 406 × 323 mm, making each folio 323 × 203 mm. The text is written on one side only. Each sheet carries a watermark of the 'Pro Patria' type, being an 'IV' on folios **30** to **112**, with the initials 'GR' surmounted by a Crown on the remaining folios. There is no date incorporated into the watermark, but when compared to contemporary dated examples the date of manufacture can be determined to between 1724 and 1726.[55]

The copyist's identity and the date when the transcription was made are not revealed. Comparing the handwriting to known types suggests that the transcripts may well have been produced by John Eckersall, who served as clerk to the Secretary of the Prince of Wales from Molyneux's appointment

53. SCA D/M/1/3, deposited by George Frederick Pitt under a deed of gift dated September 1831 (ref. SC4/2/395). The archive catalogue states that there is a microfilm master negative at the National Library of Ireland (nn. 2680–1 and pp. 1586–7) and at Cambridge University Library.

54. Edward Chamberlayne, *Angliae Notitia, or the Present State of England* (London, 1700), p. 172.

55. Edward Heawood, *Watermarks: Monumenta Chartae Papyraceae Historiam Illustrantia* (Hilversham, 1950), p. 146 and pls 491–6.

in 1714 until 1727 when the Prince acceded to the throne.[56] He was a signatory to Molyneux's will dated 15 January 1717 and after his master's death served as Receiver General and cashier of customs revenues, and as Secretary, Keeper of Seals, and Master of Requests to Queen Caroline.[57] Several blank spaces have been left in the course of transcription, which suggests that some trouble was had in deciphering or clarifying some of the words. Moreover, folios **30** to **43** carry some corrections to grammar and phraseology in what appears to be Molyneux's known hand, which suggests that the manuscript was produced in Molyneux's lifetime. This hand, compared to that in his letter-book which dates from between 1707 and 1709, is again evident on the reverse of folio **36** (St Paul's Cathedral) when the comment is written, 'Severall half burnt Stags & Roman vessells & instruments used in sacrificing to that goddess have been found here in digging the foundations of the church'. There are no alterations or corrections made after folio **50**.[58]

Each letter carries the formal address 'Dear Sir' and, as would be expected in a copy-book, ends without a signature. The recipient is not mentioned by name but may be identified with some confidence. The content of some of the letters shows that Molyneux's reader was familiar with Dublin and with Trinity College in particular. His expressed hope that 'I may give you some pleasure in remembering what you have so often seen and been so well acquainted with some time ago' and the reference to 'the young age at which you were settled here' (f. **33**) shows that the recipient once lived in London. (This would eliminate Molyneux's great friend, George Berkeley, who did not go to England until January 1713.) Furthermore, when at the Tower of London, Molyneux wrote, 'it will be needless to tell you that this is a kind of Ancient Gothick Citadell' (f. **59**), which suggests that that particular site was known to his reader. Indeed, all the evidence supports the likelihood that he was writing to his uncle, Thomas Molyneux, who at the time of his own election as Fellow of the Royal Society was only twenty-five years old and resided at The Flower de Luce near St Dunstan's Church in Fleet Street. He later worked at the Royal Mint within the walls of the Tower of London. Confirmation of this theory comes with Molyneux's visit to 'the Museum of Dr Woodward' in February 1713 (ff. **93–7**). Mocking Woodward's self-belief, he writes, 'I must observe to you that he assur'd us with the same Confidence that your great Moose deers Horns found in Ireland were transported thence by the Deluge from America' (f. **96**), a clear

56. Eckersell's style can be seen in Molyneux's 1714 to 1718 account books at SCA D/M 3/1.
57. The National Archives (hereafter TNA) Probate 11/621.
58. SCA D/M 1/2. At the end of the Dublin Philosophical Society correspondence the hand changes when George Berkeley's letters to Thomas Molyneux (May to December 1709) are transcribed.

reference to his uncle's paper on the subject published in 1695 and read to the Royal Society in London in April 1697.[59] From a young age Molyneux was in the habit of writing to his guardian uncle with accounts of his travels, as, for example, his 1709 journey into Connaught.[60] To continue corresponding in this way is not surprising, and having spent £5,000 in 1711 on setting up an asylum for the blind on Peter Street, Dublin, Sir Thomas would have been committed to his new project and may well have welcomed the diversion presented by his nephew's letters.

Little is known of Molyneux's research methods other than Thomas Hearne's recollection that 'all along that he took notes of what I said; which I construed no other-wise than as if it had been with a Design to satisfy his Curiosity, and with an intent to make a private Use of them when he had left us, not imagining that he had proposed to himself any wicked End'.[61] Evidence that Molyneux used these notes when writing his letters is suggested by the 'occasional' or phonetic spelling of some of his words, examples being 'sharawaggi' or 'sharawadgi' (a word first used by Sir William Temple (1628–99) in 1692 to describe the Chinese style of garden planting) written by Molyneux as 'scaravagie' (f. **89**) and the printers 'Bogligglilo', most likely Antonio Pollaiulo (1433–98), and 'Teillier', Johannes Teyler (1648–*c.* 1709) (f. **89**), listed whilst visiting Lord Pembroke's collections.[62]

Note on the Transcription

This transcription was made by Miriam Phillips, a volunteer in Southampton City Archives, following an initial request made by the author. The original transcription has been checked by the author. Editorial intervention has been kept to a minimum; however, some corrections have been made where appropriate and punctuation has been silently added, in both cases only where there is no ambiguity. Square brackets indicate the loss of manuscript, missing words or torn manuscript. Furthermore, for ease of referencing, the numerical sequence of folios is given in bold in square brackets. The transcription follows the original manuscript, hence any peculiarities of capitalisation and spelling are from source.

59. Thomas Molyneux, 'Discourse on the large horns frequently found underground in Ireland believing that they belonged to the American deer called Moose', *Philosophical Transactions of the Royal Society*, 19 (1695), pp. 489–512.
60. SCA D/M 1/1. Thomas Molyneux also received letters from George Berkeley in August 1709, referring to the Monastery of Buttefont and the Castle of Liscarol.
61. Rannie, *Hearne*, p. 109.
62. William Matthews, 'Some Eighteenth-Century Phonetic Spellings', *The Review of English Studies*, 12, no. 45 (January 1936), pp. 42–60.

LETTER 1

(fragment: ff. **30–2**, undated)

SCA D/M 1/3

This undated fragment (Fig. 3) describes Molyneux's journey from Dublin to either Holyhead[63] or Chester and thence to London, from where the letter appears to have been written. Molyneux mentions the 'wet season' (f. **31**) which, given our understanding of his movements, would indicate that his date of travel was October 1712. Furthermore, by referring to his passage as 'quick' (f. **32**), a hurried journey is implied rather than a prolonged topographical tour. As such, his observations during his journey tend to be simple vignettes relating to travel, farming, roads and accommodation which, at times, he compares to Irish equivalents.

Some indication of the route he may have taken is implied when he compares the poor state of the roads on his way to Oxford in February 1713 with a section of road called Chalk Hill near Dunstable in Bedfordshire, which he recalled from his earlier journey (f. **112**). On the main route between London and Chester, Dunstable is situated on Watling Street, the old Roman road that connected North Wales with Shrewsbury, Tamworth and London. Following such ancient roads often gave purpose to an antiquarian's journey, though the loss of Roman remains during surface repairs was considered lamentable. Regarding the road surface, Molyneux considered it to be 'terribly out of Repair' (f. **31**) and consequently approved of the new Turnpike Trusts set up by an Act of Parliament in 1707 that shifted responsibility for repairs from the local parish to a more effective private toll-based trust. The first turnpike was set up between Fornhill in Bedfordshire and Stony Stratford in Buckinghamshire, which may well be the section of road to which Molyneux refers.

Fragment text of letter 1

[start of manuscript]

> [**30**] [...] slight for any use but to be lay'd on the Horse that bor[e] it. Here I was surpriz'd with their moveing Mountains of Wagons and Carts, some with an Axell tree to be drawn by two Oxen, some with four Wheells, of may [be] 14 or 15 foot long and proportionable Height and Breadth with

63. Markham, *Loveday*, p. 122. Loveday wrote, 'The weather was so calm that the packet boat took twenty-six hours to cross the Irish Sea'.

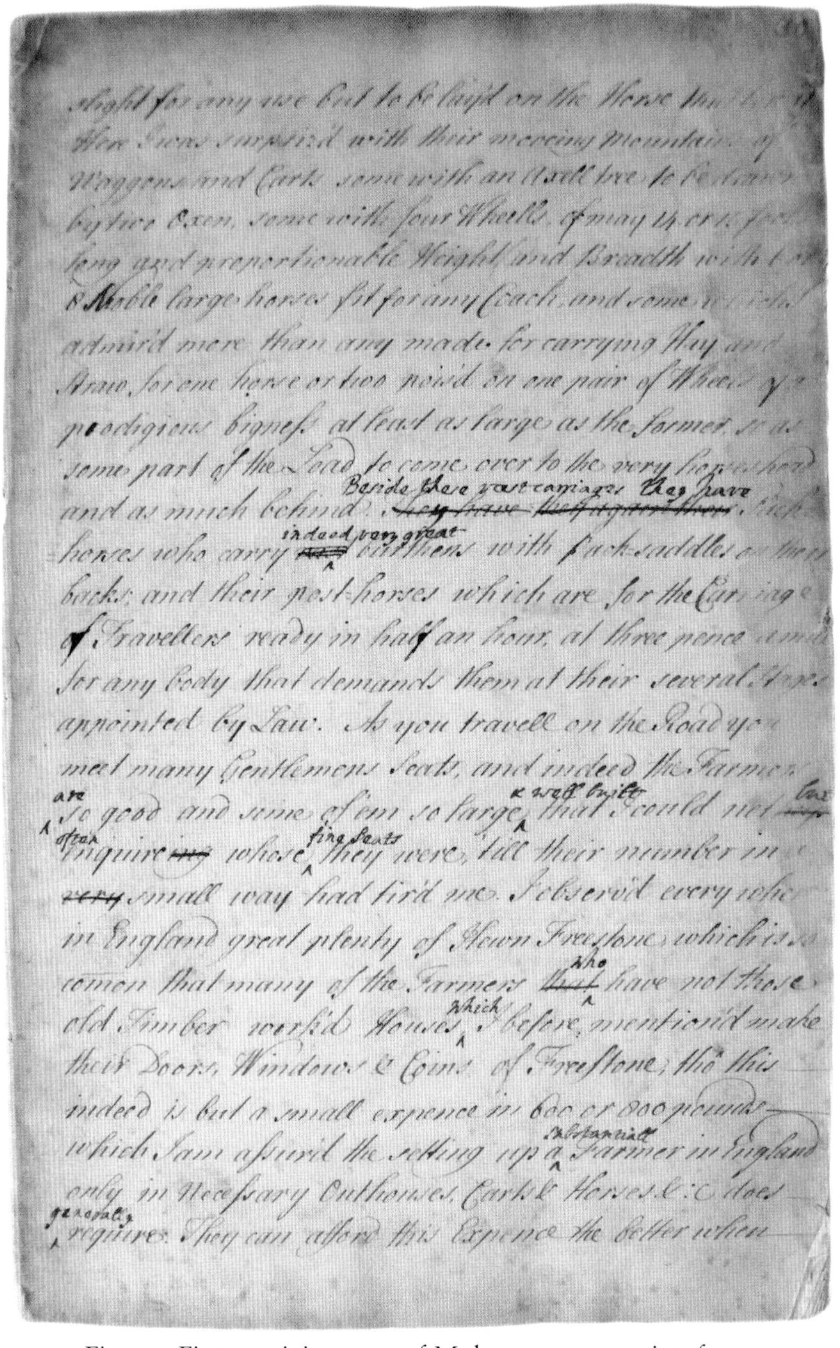

slight for any one but to be layd on the Horse ...
their Eyes surpriz'd with their moving Mountains of
Waggons and Carts some with an Axell tree to be drawn
by two Oxen, some with four Wheels of may 14 or 15 feet
long and proportionable Height and Breadth with 6 or
8 Noble large horses fit for any Coach, and some ...
admir'd more than any made for carrying Hay and
Straw, for one horse or two joyn'd on one pair of Wheels of a
prodigious bigness at least as large as the former so as
some part of the Load to come over to the very horses head
and as much behind. Beside these vast carriages they have such
horses who carry indeed very great burthens with pack-saddles on their
backs; and their post-horses which are for the Carriage
of Travellers ready in half an hour at three pence a mile
for any body that demands them at their several Stages
appointed by Law. As you travell on the Road you
meet many Gentlemens Seats, and indeed the Farmers
are so good and some of em so large & well built that I could not but
often Enquire whose fine Seats they were, till their number in a
small way had tir'd me. I observ'd every where
in England great plenty of Hewn Freestone, which is so
comon that many of the Farmers who have not those
old Timber work'd Houses which I before mentioned make
their Doors, Windows & Coins of Freestone, tho' this
indeed is but a small expence in 600 or 800 pounds
which I am assur'd the setting up a substantiall Farmer in England
only in Necessary Outhouses, Cattle Horses &c does
generally require. They can afford this Expence the better when

Fig. 3. First surviving page of Molyneux manuscript, f. 30.
The corrections appear to be in Molyneux's hand.

6 or 8 Noble large horses fit for any Coach, and some which I admir'd more than any, made for carrying Hay and Straw, for one horse or two pois'd on one pair of Wheels of a prodigious bigness at least as large as the former, so as some part of the Load to come over to the very horses head and as much behind. Beside their vast carriages they have Pack horses who carry indeed very great burthens with pack saddles on their backs, and their post-horses which are for the Carriage of Travellers ready in half an hour, at three pence a mile for any body that demands them at their several Stages appointed by Law. As you travell on the Road you meet many Gentlemens Seats, and indeed the Farmers [farmhouses] are so good and some of 'em so large and well built that I could not but often enquire whose fine Seats they were, 'till their number in a small way had tir'd me. I observ'd everywhere in England great plenty of Hewn Freestone which is so comon that many of the Farmers who have not those old Timber work'd Houses which I before mention'd make their Doors, Windows and Coins of Freestone, tho' this indeed is but a small expence in 600 or 800 pounds which I am assur'd the setting up a substantial Farmer in England only in Necessary Outhouses, Carts and Horses &:c does generally require. They can afford this Expence the better when [**31**] generally they hold their Lands as Freeholds so that they improve for their own posterity, and have such a constant sure Market by the number of their People that they raise great Fortunes and live very well by liveing industriously; insomuch that you shall meet a right pretty young Maid rideing to Market in her Straw Hat worth may be a thousand pounds to her portion, these Straw Hatts by the way are in my opinion very becomeing and a very proper dress for the Country, and very Cheap, one of them lin'd with Silk not standing them in above half a Crown. The Country of England is very remarkably plain all the way from Chester, which with the goodness of their Ground and the Weight of their Carriages makes the roads most insupportably bad in this wet season of ye year and indeed I have been wonderfully surpriz'd to find that in so rich and tradeing a Country as England their Roads should be so terribly out of Repair as they are, it is at least one good which we receive by Our weak horses in Ireland that we are forc'd to have good roads for them: I observ'd however in two places between London and Chester the road had been for a Mile or two lately repair'd by authority of Parliament and here we had a gate Cross the road, and pay'd a penny a horse for Our Passage which is ye fund for these reparations of ye road. If their ways are much Worse than ours they exceed us in one improvement which I observ'd here and there and that is they have very frequently at severall ways meeting a Post set up with several hands and Indexes fixt on it to direct a Traveller which way each road leads, together with the distance of the next Town [**32**] this is a great improvement of the Roman Stones and would very well be introduc'd into Ireland to the great comfort of a Traveller in a Country where it may be he has no one in a Mile to ask a question of and those he meets often don't know his language or can't

tell their Right hand from their left. I observ'd nothing remarkeable in Our Dyet on the Road but that we had almost every where excellent water and as good bread, Our Wine very ordinary, and seldom French Claret which the War has almost banish'd. On the road we burnt wood every where which shews its great plenty and that I may omit nothing to the Honour of Ireland had mighty bad Linnen in every house. These are some of the general remarks I made in passing thro England, if you don't find them just I can only say I pass'd so quick thro' the Country that I could not make 'em more exact and such as they are if they serve to divert you for half an hour I have my end and am Yrs &:c.

LETTER 2

(three fragments: ff. **33–43, 58–63, 77–82**), 20 December 1712

SCA D/M 1/3

Assuming that the first letter dates to October 1712, between seven and ten weeks have passed before Molyneux fulfils his promise of writing home. For this oversight he blames 'The vast variety of diverting objects which occur in this great City' (f. **33**). Between October and 20 December 1712 Samuel Molyneux sets off from his rented rooms in Suffolk Street (f. **151**) (Map I ①) on tours of the foremost attractions in his immediate area. In this letter, the only one that refers solely to the curiosities within, or in close proximity to, the city of London, he describes his visits to St Paul's Cathedral, Westminster Abbey and the Palace of Westminster, before a section of the manuscript is lost after folio **43**. When it resumes at folio **58**, Molyneux is visiting an unknown library. He goes on to provide a comprehensive description of the Tower of London until folio **63**, after which again the manuscript is lost. From folio **77**, the last section of the letter, he mentions viewing a number of scientific instruments devised by members of the Royal Society and visiting two notable private collections, one belonging to John Kempe and the other to Thomas Herbert, the 8th Earl of Pembroke.

Primarily a natural philosopher, Molyneux's confidence and enthusiasm in his correspondence are greater when writing as a scientist rather than an antiquarian. His election as a Fellow of the Royal Society demonstrates his fervour and proficiency in this subject. During the first decade of the eighteenth century the Royal Society tended to shun the study of antiquarianism, prompting the creation in 1707 of a new learned body — the Society of Antiquaries of London. Although never a member himself, Molyneux's search for self-improvement and edification reflects the society's *raison d'être*.[64] Being new to London, it is unsurprising that he describes himself at the outset of his correspondence as 'so ignorant a Traveller and perfect Stranger' (f. **33**). Of his purpose, he affirms his intention to his reader to '[...] give you some pleasure in remembering what you have so often seen and been so well acquainted with some time ago' (f. **33**), a feat he achieves by being conscious of his surroundings, earnest in description and modest of opinion. His energetic accounts, flitting between antiquity and science, in this second letter set the tone for the rest.

64. Dai Morgan Evans, 'The Society of Antiquaries, 1707–18: Meeting Places and Origin Stories', *The Antiquaries Journal*, 89 (2009), pp. 323–35.

For Molyneux, intellectual exchange was a vital component in his quest for knowledge and, as the correspondence indicates, during his time in England he was inclined towards a wider sociability. At St Paul's Cathedral (Fig. 4) (Map III ②), for example, he mentions touring with 'company', and also records a direct conversation with its architect, Sir Christopher Wren (f. **36**). Perhaps influenced by their meeting, Molyneux touches on two of Wren's primary concerns about his masterpiece. With regard to the interiors, Molyneux remains undecided on the merits of an undecorated dome (an issue Wren had raised with Queen Anne), perhaps influenced by the Low Church notion that religious images were inappropriate symbols of Popery and therefore inconsistent with the homilies of the Church of England. Indeed, the Whig Bishops were so outraged that Wren, a secular architect, was attempting to set a religious agenda that a competition was set up in 1709 to prevent Louis Laguerre (1663–1721), Wren's choice of artist, from creating an iconographic depiction of the life of St Paul on the dome. Secondly, despite being admired by Molyneux (f. **36**), 'the Iron rails round the Church yard' were judged by Wren to be 'ugly, extravagant and quite unsuitable' — clearly a sore point with the architect whose own design for a wrought-iron churchyard fence had not been considered by Dean Godolphin and the Building Committee.

Work on the cathedral had started in 1675 and was declared finished by Parliament in 1711 after £736,752 had been spent on its construction.[65] Molyneux's Uncle Thomas, therefore, would have been resident in London during the early stages but would not have witnessed its completion, hence the reason for such a precise and detailed description. Molyneux's quasi-theological narrative portrays the cathedral as much an anti-climax as it is a heavenly pleasure. In contemplation of Wren's accomplishment, he does not wholly approve of its situation, style and proportion, a view which several contemporaries shared and which Daniel Defoe later condemned when he wrote that 'some authors are pleased to expose their ignorance, by pretending to find fault in it'.[66] By contrast with the magnificence of the main body of St Paul's, Molyneux expresses disappointment in the choir and its 'mean' altar-piece, but compliments Francis Bird's fine carving and sculpture of Queen Anne and the design of the cupola. In the Model or Trophy Room situated above the Consistory Court (since 1901 the Chapel of St Michael and St George) in the north aisle, Molyneux examines two models of the cathedral. The first, made of oak and plaster by William Cheere (d. 1690) and completed by August 1674 (although abandoned soon after), is described as being '10, or 12 feet long' (f. **39**), although in actuality it measured 6.41 metres (21 ft) long and 4 metres

65. R. and J. Dodsley, *London and its Environs Described* (London, 1761), p. 160.
66. Furbank, Owens and Coulson, *Defoe Tour*, pp. 142–3; see also James Ralph, *A Critical Review of the Publick Buildings, Statues and Ornaments in, and about London and Westminster* (London, 1734), pp. 18–23.

Fig. 4. 'The North Prospect of St Paul's Cathedral in London', by Thomas Bowles, *c.* 1745.
© *NTPL/Derrick E. Witty*

Fig. 5. St Paul's Cathedral, the Great Model.
Photograph with kind permission of Mr Sampson Lowe

(13 ft) high. The second 'somewhat broken' model (Fig. 5) alludes to Wren's early Greek cross design to which Robert Hooke referred in 1672 when he wrote of the 'model [...] approved by the King'.[67] Interestingly, Molyneux records the architect's intention that this room and the corresponding chamber above the Morning Chapel in the south aisle (since 1905 the Chapel of St Dunstan) are both destined to be libraries, but does not mention either the majesty of the architectural space or the books that were already *in situ* by 1707.

Following his methodical account of St Paul's Cathedral, less attention is paid to the merits of Westminster Abbey (Fig. 6) (Map VI ③) which, for a building that effectively embodies English history, is surprising. Such brevity may suggest a fleeting visit or perhaps reflects some disappointment in a building that Defoe later described as 'the ruins of a church'. Neglecting the architectural and monumental details and the programme of repairs to the western and crossing towers being carried out under the supervision of Sir Christopher Wren as Surveyor of the Fabric, Molyneux occupies himself in pedantic manner with a meditative walk around the monuments and tombs (ff. **39–40**), echoing Joseph Addison's essay from *The Spectator:*

67. For more on the Great model, see Kerry Downes, *Sir Christopher Wren and the Making of St Paul's* (Royal Academy of Arts, 1991), pp. 10–14 and 52–3.

Fig. 6. Westminster Abbey, interior view of King Henry VII's chapel, *c.* 1710.

> When I am in a serious Humour, I very often walk by my self in *Westminster- Abbey*; where the Gloominess of the Place, and the Use to which it is applied, with the Solemnity of the Building, and the Condition of the People who lie in it, are apt to fill the Mind with a kind of Melancholy, or rather, Thoughtfulness, that is not disagreeable'.[68]

As at St Paul's Cathedral, Molyneux is guided around the Abbey. However, taking his misinformed guide at face value, he refers incorrectly in his letter to the continued existence of Edward the Confessor's regalia and jewels, the most precious of which (the crown, sceptre and tunic once used for coronations) had been destroyed by Cromwell's men (f. **39**).

As a stranger in London, Molyneux would have undoubtedly taken information from written sources. Never acknowledging any sources explicitly by name, some of his rhetoric is reminiscent of that found in Edward Hatton's two-volume publication *New View of London* (1708) — a source used extensively by von Uffenbach during his visit to London only two years before Molyneux was there.[69] Copies of *New View of London* were held in Molyneux's library at the time of his death but, despite being considered inaccurate by serious eighteenth-century antiquaries and historians, Molyneux's account of Westminster Palace (Map VI ④) repeats, in part, Hatton's own observations.[70] For example, when Hatton claims that Westminster Hall (Figs 7 and 8) is 'probably the most capacious room in Christendom', Molyneux is only slightly more judicious when he describes it as 'one of the noblest Largest single Rooms that is perhaps in all the World' (f. **40**). Neither pays much attention to its architectural merits, the interiors or its functional spaces such as the law courts and shops, which Hatton simply describes as comprising '[...] chiefly booksellers and milliners'. Furthermore, both insist inaccurately that Irish oak was used for Westminster Hall's spectacular hammer beam roof — a point later challenged by Don Manoel Gonzales when he wrote 'this is a fact not to be depended on, for I find that the timber brought for rebuilding and repairing the Palace of Westminster in the reign of Richard III was brought from the forests in Essex'.[71]

Molyneux's description of the Palace is abstracted and hurried. Passing through the Houses of Commons and Lords, then situated in St Stephen's chapel (Fig. 9) and the parliament chamber, with the painted chamber alongside, only a brief mention is made of the Spanish Armada tapestries in the Lords chambers (f. **42**), while the Trojan War tapestries in

68. *The Spectator*, No. 26, 30 March 1711, pp. 103–4.
69. Bridget Cherry, 'Edward Hatton's "New View of London"', *Society of Architectural Historians of Great Britain*, 22 (2001), pp. 96–105.
70. Edward Hatton, *New View of London*, II (London, 1708), p. 637.
71. Gonzales, *London in 1731*, p. 32.

Fig. 7. Westminster Hall, view of the exterior, *c.* 1700.
© *The Trustees of the British Museum*

the painted chamber, described by von Uffenbach as 'uncommonly ugly and worn',[72] are not mentioned at all. Perhaps the Palace was merely a distraction from his eagerness to visit the 'handsome house' (Map VI ⑤) of fellow Whig Charles Montagu, then Baron Halifax (1661–1715), one of the foremost connoisseurs of his age. Halifax collected actively throughout his life and in October 1712 had commissioned John Talman to inspect the Duke of Mantua's collection of paintings, books, manuscripts, statuary and busts with a view to purchase — although it would appear that he was unsuccessful in acquiring any of the items.[73]

72. Quarrell and Mare, *von Uffenbach*, p. 74.
73. Cinzia Maria Sicca, 'The Making and Unravelling of John Talman's Collection of Drawings', in Cinzia Maria Sicca (ed.) , *John Talman; an Early-Eighteenth-Century Connoisseur*, Paul Mellon Studies in British Art 19 (New Haven and London, 2008), pp. 45–6. The sculpture had been acquired by the Venetian collector Zaccaria Sagredo.

Fig. 8. Westminster Hall, "The First Day of Term" Print by Charles Mosley, 1797.
© *Palace of Westminster Collection.*

Access to Halifax's collection, it would seem, was based on good letters of introduction and mutual Royal Society connections; indeed, Halifax was President of the Royal Society between 1695 and 1698 and therefore would have known Molyneux's father and uncle. It is certain that Molyneux would have been guided personally around these apartments, perhaps by the Mr Dale referred to later in the manuscript (f. **58**) or even by Halifax himself, who is listed later in the manuscript as '[one of those] I have had the Honour since I have been here to be sometimes receiv'd by' (f. **148**). However, a personal meeting with him is not specifically mentioned in the surviving sections of manuscript so, if not at Westminster, the two men may have become acquainted a little later when Molyneux visited Bushy Park in Middlesex (f. **86**), where Halifax was Ranger.

As Auditor of the Exchequer (a life-time appointment), Halifax was provided with grace-and-favour apartments alongside the eastern and northern elevations of the medieval cloisters overlooking the River Thames.

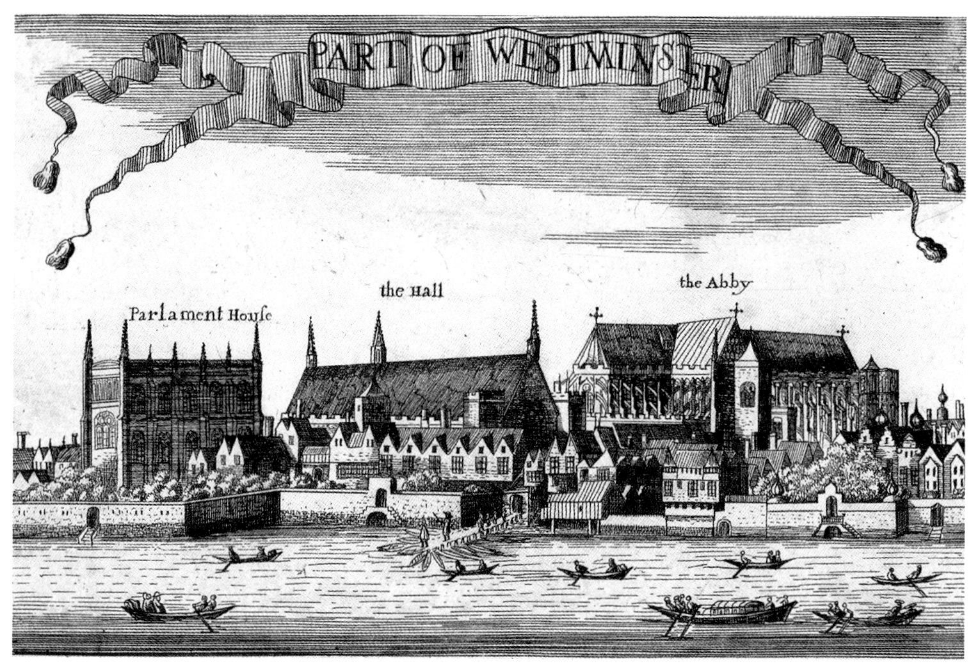

Fig. 9. Westminster from the river, *c.* 1690.
© *The Trustees of the British Museum*

Molyneux refers to the Baron's 'three noble large rooms' (f. **43**), which are shown on a later plan by R. W. Billings, one measuring 30 by 25 feet and the other two measuring 20 by 25 feet (one with a large, bayed, east-facing window overlooking the river).[74] William Nicolson, Bishop of Carlisle, had vouched for the richness of the collection when he wrote in his diary on 13 January 1705:

> N.B. In the way to the House, I went (with Mr Holmes and Mr Dale) to my Lord Hallifax; who kindly carried me into his Library, a gallery nobly furnished with curious Books placed under statues as in Cotton's. His Lordship has Transcripts of all the Rolls of Parliament, Journals of the Lords and Commons, &c. and had interleaved Thuanus's Catalogue for shewing the Rarities and Defects of his own collection.[75]

Molyneux concurs, describing some of the statuary and pictures and briefly mentioning the library, in which he focuses his attention on the curious system of shelving (f. **43**) just before the manuscript breaks off.

74. Maurice Hastings, *Parliament House* (London, 1950), p. 119.
75. Clyve Jones and Geoffrey Holmes (eds.), *The London Diaries of William Nicolson, Bishop of Carlisle, 1702–1718* (Oxford, 1985), p. 277; *Catalogus bibliothecae Thuanae* (Paris, 1679) by Jacques-Auguste de Thou ('Thuanus'), the French historian and statesman (1553–1617).

The Bishop of Carlisle's companions in 1705 were George Holmes (1662–1749), Deputy Keeper of Records at the Tower of London from 1690 and an early member of the Society of Antiquaries of London, and, most likely, Robert Dale (1666–1722), Clerk of Records at the Tower and later Richmond Herald at the College of Arms, and also a great friend and companion of the diarist Ralph Thoresby. Molyneux appears not to have met Holmes or Dale during his visit to the Tower of London, although it seems unlikely that he would have missed the chance to see its renowned manuscript collections, so a visit may well have featured amongst the missing folios later in Letter 2 (ff. **64–76**). Yet when the manuscript resumes, approximately 3,500 missing words later at folio **58**, Molyneux is being shown a collection of manuscripts and journals by Mr Dale. Whether this is the same Mr Dale referred to by the Bishop of Carlisle remains unclear, as here Molyneux portrays Dale as a 'Clothier' (f. **58**), working 'near the Exchange'. Other antiquarians known to have come from humble backgrounds include John Stow (1525–1605), a tailor who published his *Survey of London* in 1598; the Harleian librarian Humfrey Wanley (1672–1726) who was the son of a draper; and the book, print and curio dealer, John Bagford (1650–1715), who was a shoemaker.[76] As a 'Clothier' at the Exchange, Dale would have benefited from the cosmopolitan milieu described by *The Spectator:*

> It gives me a secret Satisfaction, and, in some measure, gratifies my Vanity, as I am an *Englishman*, to see so rich an Assembly of Countrymen and Foreigners consulting together upon the private Business of Mankind, and making this Metropolis a kind of *Emporium* for the whole Earth. [...] Sometimes I am justled among a Body of *Armenians*: Sometimes I am lost in a Crowd of *Jews*: and sometimes make one in a group of *Dutchmen*. I am a Dane, Swede or Frenchman at different times; or rather fancy my self like the old Philosopher, who upon being asked what Countryman he was, replied, That he was a Citizen of the World.[77]

The missing folios and interrupted text make it unclear where Molyneux is on folio **58**. It is possible that he is still in Halifax's apartments and that the collection of parliamentary books and journals he is writing about is the same as that seen by Bishop Nicolson; however, the estimated number of missing words, twice the length of the full account of the Harleian collection and three times that of the Bodleian library in Oxford, makes this unlikely. It is tempting to suggest that Molyneux and Dale are in the Cotton Library which was, as John Strype explained in 1720, 'In this Passage out of

76. Rosemary Sweet, *Antiquaries: The Discovery of the Past in Eighteenth-Century Britain* (London, 2004), p. 58.
77. *The Spectator*, No. 69, 19 May 1711, pp. 268–9.

Westminster Hall into the old Palace Yard, a little beyond the Stairs going up to St. Stephen's Chapel, (now the Parliament House) on the left Han'd'.[78] We know that Molyneux does visit the Cotton Library, as he mentions it later on folio **126**, but again no actual description of his visit has survived. However, this suggestion is problematic as Molyneux's description of a house 'peculiarly neat and elegant in all respects' (f. **58**) does not correspond with the same Cotton House described by the *House of Lords Journal* as damp and in a 'ruinous Condition', threatened with demolition in 1707 and a cause of such concern that in 1716 a petition was presented to the Treasury for its repair.[79] It would also be more likely that if he was visiting the Cotton Library Molyneux would expect to be shown round not by Mr Dale but by the Cotton Librarian, William Hanbury, or by his assistant John Elphinstone.[80] Both John Elphinstone and George Holmes were later known to Molyneux when in 1715, as Secretary to the Prince of Wales, he drew on their services in transcribing documents from the Cotton Library relating to Chester, Cornwall and Wales, followed a year later by 'papers out of the Tower'.[81] Another suggestion is that Molyneux was visiting Ashburnham House in Little Dean's Yard, Westminster (now part of Westminster School). Built for William Ashburnham (d. 1679) in *c.* 1662 probably by the architect John Webb (most likely to the designs of Inigo Jones), in 1712 the house was owned by John (1687–1737), 3rd Baron Ashburnham and would certainly befit Molyneux's description of 'neat' and 'elegant' (f. **58**). Mr Dale may well have been his librarian.

Molyneux's visit to the Tower of London (Map IV ⑥), attending seemingly as a regular tourist despite his personal contacts, is a more straightforward account. His protracted description of the collections appears peculiar at first, considering his uncle's previous knowledge of the site. For the modern reader, though, such a comprehensive record of the armouries and crown jewels helps to bridge a gap between early visitors' accounts and published guides, such as *The Antiquities of London and Westminster* (1722). Arriving by water at the St Thomas's Tower (Traitors' Gate) entrance, Molyneux immediately examines the foundations of the Roman wall and bastion on which the Wardrobe Tower was later built. As

78. John Strype, *A Survey of the Cities of London and Westminster* (2 vols, London, 1720), II, 55. The collection was put together by Sir Robert Cotton, 1st Baronet (1571–1631), antiquary and politician, and included books and manuscripts given to Cotton by fellow antiquary William Camden (1551–1623). It was sold by Sir John Cotton, 3rd Baronet (1621–1702), in 1701.
79. *House of Lords Journal 1705–1709*, 18 (March 1706), pp. 156–7; William A. Shaw and F. H. Slingsby, *Calendar of Treasury Papers 1716*, 30 (London, 1958), p. 213.
80. Colin G. C. Tite, *The Manuscript Library of Sir Robert Cotton*, The Panizzi Lectures, 1993 (London, 1994).
81. SCA D/M 3/1 and 4/2–5.

a rule, antiquarians tend to favour the past with more detail than the present and to distance themselves from overtly political statements — at the Tower, Molyneux does neither. In a similar vein to when he sees the patriotic display of flags in Westminster Hall (f. **42**), on sight of the Roman remains and the military spoils taken from the French during the War of the Spanish Succession (f. **59**), Molyneux's thoughts turn to what he regards as the morally justifiable war abroad and to the welfare of the recently exiled Duke of Marlborough. At a time when the truce already agreed to end the European war made the forthcoming Treaty of Utrecht a formality, and while the Whig party was in so much disarray, such partisan views could be considered dangerous and, as such, sit uncomfortably in the narrative.

As a self-proclaimed 'man of good nature & benevolence to mankind', Molyneux finds it 'impossible [...] to walk with Pleasure in this Repository of Death & Destruction' (f. **61**). Despite such high-minded sentiments, it seems that his tour of the Armoury in the two-storey Great or Grand Storehouse,[82] with its impressive displays of foot armoury and heavy artillery, appeals to his notion of national identity. Such pride strengthens when he explores the collections with cultural or 'ornamental' significance (f. **61**), the Horse Armoury (often referred to as 'the Line of Kings') located on the first floor of the New Armouries to the south-east of the White Tower and the Spanish Armoury on the first floor of a storehouse north of the Wakefield Tower.[83]

Molyneux's account of seeing the Crown Jewels in the Martin Tower (or Jewel Tower) echoes that of other eighteenth-century travellers, such as von Uffenbach who in 1710 calls the room 'a gloomy and cramped den', and William Hunton who, much later in 1784, writes of a 'dismal hole resembling the cell of the condemned'.[84] The inadequate care given to such a highly prized collection causes him to speculate on the authenticity of the Crown Jewels and of their security, although curiously he makes no reference to Captain Blood's failed attempt to steal part of the collection in 1671. His sometimes fallacious comments may reflect some annoyance at not receiving any special privileges; indeed, he has to pay to see the Coronation regalia, a 'Custom' that later causes him 'great Offence' (f. **131**) when visiting the Oxford colleges.

Two themes that resonate in Molyneux's account of the Tower are defence and ornament (f. **62**). His final sentence before the manuscript breaks off again at folio **63** gives some indication of a third.

82. Built between 1688 and 1691 most likely by Sir Thomas Fitch or by an Ordnance engineer (although often attributed to Sir Christopher Wren).
83. N. Blakiston, 'The Great Storehouse in the Tower', *Architectural Review*, Volume 121 (June 1957), p. 453.
84. Quarrell and Mare, *von Uffenbach*, p. 38.

> I remember to have heard of a King that us'd to say in Mirth that Man could do any thing that had his Sword in his hand, a Crown on his head and money in his Pocketts; We have arm'd and crown'd Brittain and before we leave the Tower we must give her the third [...]

This third theme, and therefore apparently part of the subject matter for the missing folios, would appear to be money: in this context, the Mint. Thomas Molyneux would have been very familiar with the Mint, having worked there as a comptroller under Isaac Newton, although his time there ended unpleasantly when in 1700 he was charged with corruption.[85] His uncle's employment at the Tower also explains Molyneux's reference to 'my Uncle Lucas [...] Governour of the Tower' (f. **63**), which refers to Thomas, 3rd Baron Lucas of Shenfield, Governor of the Tower between 1688 and 1702. Speculating further on the missing folios, it would be surprising if he had not visited the menagerie, then the Tower's top attraction, or taken the opportunity to look at the ancient records in the White Tower.

Another key event that fits chronologically with the period covered by this letter, and therefore most likely to form part of the missing content, is the occasion of Samuel Molyneux's election as a Fellow of the Royal Society (Map II⑦) on 1 December 1712. When the manuscript rejoins at folio **77**, Molyneux appears to be at the Royal Society, which by the end of 1712 had moved into new premises at Crane Court in Fleet Street, described in an early guidebook as 'The Repository of Curiosities is a Theatrical Building, resembling that of Leyden in Holland'.[86] Some text is torn away at this point, but it is clear that Molyneux is describing a barometer and air pumps. His guide during the visit is the influential experimental physicist Francis Hauksbee (1666–1713), who from 1704 was the 'demonstrator or curator of experiments' as well as being an expert himself in the workings of air pumps and the designer of six known types.[87] Hauksbee was well acquainted with both William and Thomas Molyneux, and corresponded with Samuel from 1707 when the latter was Secretary of the Dublin Philosophical Society.[88] This account is, most likely, the last record of a first-hand encounter with Hauksbee and his work, since his final appearance at the Royal Society took place in January 1713.

At the bottom of folio **77** (Fig. 10, p. 48) Molyneux draws three concentric circles, each including inserted text reading:

85. Richard S. Westfall, *Never at Rest: A Biography of Isaac Newton* (Cambridge, 1980), p. 613.
86. Quoted in Lisa Jardine, *On a Grander Scale: The Outstanding Career of Sir Christopher Wren* (London, 2002), pp. 437–8.
87. Gillispie, *Scientific Biography*, VI, pp. 169–75.
88. SCA D/M 1/2.

centre — 'Center Pasteboard unpainted'
middle — 'A dirty white painted'
outer — 'Different colours in such proportions as Newton directs in radius as represented here'.

Accompanying the drawing, the only illustration in the whole manuscript, is the inscription 'a painted pastboard of the annex'd figure which by wheeling round gave a sensible Demonstration of what Sir Isaac Newton advances of the Composition of white in his Book of Colours' (f. **77**). Hauksbee had been a great influence on Newton, and it would appear that the pasteboard experiment could have been an indication of the revisions Newton intended to make to the forthcoming second English edition of his book *Opticks* (London, 1704), although it was not republished until 1717. Perhaps still in Hauksbee's company, Molyneux visits 'Willsons, a glass grinder' (f. **77**), most likely the James Wilson who is known to have worked at The Willow Tree, Cross Street (Map II⑧), Hatton Garden, between 1702 and 1710.[89] Primarily influenced by his father's published research in this branch of physics, Molyneux's interest in lenses and glass grinding was enduring, having a chapter published posthumously on the subject in Robert Smith's book *A Compleat System of Optics* (1738).[90]

Like many early eighteenth-century antiquarians and philosophers, Molyneux shows a great interest in coins and medals, which were deemed as reliable and truthful representations of the past. Two of the most notable collectors of the day were John Kempt or Kempe (1665–1717) and Thomas Herbert, the 8th Earl of Pembroke (1656–1733). Judging by the length and attention to detail in these accounts of such notable collections, accuracy was paramount in maintaining his reputation as a serious scholar. Indeed, of John Kempe's antiquarian collections, Bishop Nicolson remarked in his diary that it was 'too valuable for a subject to keep in his own private collection', while a later visitor, John Ward, thought the collection was simply unequalled.[91] Kempe's collection, housed in his museum 'in the lower end of the Hay-Market (Map I⑨), and the corner of Pall-Mal[l], St. James's Pall Mall', was acquired from the connoisseur George, 1st Lord Carteret, who in turn had collected from John Finch, the 3rd Earl of Winchelsea, and the antiquarians John Gaillard of Angers and Dr Jacob Spon. Molyneux's

89. *Philosophical Transactions*, 23 (1702–3), pp. 1241–7.
90. William Molyneux, *Dioptrica Nova, A treaties of dioptricks in two parts, wherein the various effects and appearances of spherick glasses, both convex and concave, single and combined, in telescopes and microscopes, together with there usefulness in many concerns of humane life, are explained* (London, 1692); Samuel Molyneux, 'The Method of Grinding and Polishing Glasses for Telescopes, Extracted from Mr Huygens and Other Authors', in Robert Smith, *A Compleat System of Optics* (Cambridge, 1738), pp. 281–301.
91. Joseph M. Levine, *Dr Woodward's Shield: History, Science, and Satire in Augustan England* (University of California Press, 1977), p. 338, n. 57.

lively commentary draws attention to the problems with forgery and unreliable provenance which vexed antiquarians throughout the Enlightenment period. Kempe, himself a skilled counterfeiter, in conversation gives the correct supposition that the embossed iron buckler in Dr Woodward's collection was 'not [...] very ancient [f. **79**] and certainly not the Roman antiquity many believed it to be'. After admiring Kempe's coins and medals, many of which derived from the collections of Thomas Howard, Earl of Arundel (1585–1646), Molyneux's attention turns to his collection of mostly Italian marbles which, at the time of Kempe's death, comprised

> [...] eleven statues [or statuettes], but almost all of them under two feet in height, besides twenty busts, sixteen reliefs and a remarkable number of inscriptions [...] The principal portion of the antiques, however, consisted of the small bronzes, among which were sixty-three statuettes, which at that time gained for the collection considerable fame.[92]

Most of the items singled out here by Molyneux are listed in the catalogue compiled after Kempe's death, and in the auction particulars of March 1721.[93] Pieces from Kempe's museum passed into the ownership of Hans Sloane and Charles Townely, and latterly into the British Museum collections.

At first sight Molyneux shows considerable expertise and discernment in antiquarian matters, although quite how much of his learning is drawn from his companion or guide, Signor Francesco Bianchini (1662–1729), we cannot be sure.[94] Bianchini was a Roman Catholic priest, archaeologist, antiquarian, astronomer and collector of antiquities. After studying in Italy, he became Gentleman of the Bedchamber to the Pope, foreign associate of the Académie Royale des Sciences and was elected as a Fellow of the Royal Society on 29 January 1713, two months after Molyneux — both men being proposed by Sir Isaac Newton, President of the Royal Society between 1703 and 1727. Referred to as 'Francesco Benedetti' by von Uffenbach, such a knowledgeable 'Gentleman' would make an ideal guide around London's antiquarian

92. Adolf Michaelis, *Ancient Marbles in Great Britain* (Cambridge, 1882), pp. 48–9.
93. Robert Ainsworth, *Monumenta Vetustatis Kempiana et vetustis scriptoribus illustrate cosque vicissim illustrantia* (London, 1720); 'A catalogue of the antiquities of the late ingenious Mr. John Kemp, F.R.S. consisting of two of the finest mummies (esteem'd so by the curious) [...] with a valuable collection of books of antiquities: will be sold at the Phoenix Tavern, next door to his museum, the lower end of the Hay-Market, and the corner of Pall-Mal [*sic*], St. James's; on Thursday the 23d of March 1721'.
94. An engraving by, or after, Pier Leone Ghezzi entitled 'Francesco Bianchini holding the eyepiece mount of an aerial telescope', dated 28 September 1720, can be found in V. Kockel and B. Solch, *Francesco Bianchini (1662–1729) und die europaische gelehrte Welt um 1700* (Berlin, 2005), p. 21.

collections and bookshops;[95] indeed, so high is Molyneux's respect for his learned Italian friend that whenever his opinion is offered it is duly recorded for the benefit of his erudite uncle. One such example appears at John Kempe's when, of the authenticity of his medal collection, 'I remember Signore Bianchini suspected grievously most of his rare ones' (f. **78**).

Like Kempe's cabinet of curiosities, part of the Earl of Pembroke's collection (Map I ⑩) of coins, medals and sculpture was acquired from the Arundelian collections. Between 1686 and 1733 the collection was housed at 12 St James's Square, where it was catalogued in the 1720s.[96] In 1696 John Evelyn remarked on the earl's 'rare Pictures of many of the old & new best Masters', while two years later the Leeds antiquary Ralph Thoresby wrote about '[the] incomparable museum of medals [where His Lordship showed] a strange variety of counterfeits [...] hard to distinguish from the originals'.[97] He added:

> His Lordship's humanity and condescension were extraordinary ; for, not thinking this kindness enough, he desired me to make him another visit, and he would show me other views, which I, modestly declining, (considering his Lordship's public station, being President of his Majesty's Privy Council, and one of the Lords Justices) but my Lord would have me promise to dine with him the next day at three, when the Council would be over, desiring, in the mean time, the perusal of my manuscript catalogue of my coins.[98]

Pembroke's collection of Renaissance and Baroque prints were also of great interest to Molyneux. Although he specifically mentions 'Bogliggliolo' (f. **81**), most likely Antonio Pallaiuolo (1433–98), a Florentine painter who made a famous early engraving known as the *Battle of the Nude Men*, and Claude Mellan's (1598–1688) print entitled *Sudarium of Saint Veronica* (1649), where the engraving comprises of a single line spiralling from the tip of Christ's nose, it is another 'by [...] Teiller, in different colours' (f. **81**) that remains intriguing. Although the first colour prints were made in 1710 by Jacob Christoph Le Blon (1667–1741), it would seem that Molyneux was referring Johannes Teyler of Haarlem who experimented with colour printing during the latter part of the seventeenth century. Of Pembroke's library, Molyneux mentions '60 [volumes] of Elzevir printing' (f. **81**), Elzevir

95. Quarrell and Mare, *von Uffenbach*, pp. 70 and 183.
96. Nicola Haym, *Numismata Antiqua Pembrochiana* (1726); the coins and medals were later kept at Wilton House and sold off in 1848. See Wiltshire and Swindon Record Office 2057/H/6/1.
97. de Beer, *John Evelyn*, p. 880; Hunter, *Ralph Thoresby*, pp. 335–9.
98. Hunter, *Ralph Thoresby*, p. 336; Jonathan Scott, *The Pleasures of Antiquity: British Collectors of Greece and Rome* (New Haven and London, 2003), pp. 39–49; Michael Hall (ed.), 'The Earl of Pembroke: Wilton and its Collections', *Apollo* (July/August 2009), pp. 36–62.

being a Dutch family of printers who produced fine editions but ceased to trade in 1712.

Letter 2 highlights the trials of many early eighteenth-century antiquarians. Devoting himself to accuracy, aesthetic appreciation and to the elucidation of numerous facts, Molyneux carefully reduces his complex thoughts and numerous experiences to the simplicity of the written word. His frustration, at times, of such an impossible task is palpable, although he does clearly present the nature of antiquarianism as a rational debate and the collections themselves as a commodity of the age. Yet, if we sympathise with Molyneux in trying to underpin his visions of the past, then things become even more difficult in articulating the present. His difficulties with St Paul's Cathedral, for example, a building that reflected the closing stages of an artistic age and thereby shows the upheavals within it, reveals Molyneux's progressive approach to art — something that becomes more apparent in the next letter.

Text of Letter 2 in three fragments

[**33**] Dr Sir, Dec: 20th 1712. London

The vast variety of diverting objects which occur in this great City, to so ignorant a Traveller and perfect a Stranger as I was have prevented me very often in my Designs of sitting down according to my promise in order to give you some Account of the most remarkeable things I have found since I have been here, I am resolv'd tho' now to take this Evenings leisure and tho I can by no meanes pretend to give you exact descriptions yet I am not without hopes I may give you some pleasure in remembring what you have so often seen and been so well acquainted with some time ago. It may not be impossible I may observe some things to you which you did not know before at least I am sure I meet several who have been born and liv'd allways here and yet know very little or nothing at all of the Curiositys of this Place, whether the young age in which you were settled here has brought upon you the same inadvertency or that your natural Curiosity hath got the better and given you to observe what was remarkeable or not I know not, for my part I think my self happy in haveing liv'd at home to a fit age for wonder and observation. I shall keep no more of Order in my Descriptions, than I did in the sight of the things themselves but as any one occurs to my memory I shall put you in mind of it. [**34**] To begin therefore with the great Cathedral Church of St Paull in a view of the Citty nothing so soon strikes ones Eye as this vast Fabrick, but especially upon the River as you pass in a boat you have a very fair prospect almost of the whole Church, not only the Cupola but also the body of it; far exceeding the height of the adjacent buildings it has something very magnificent and noble both in its Size and Structure, it is built in the Antique Roman manner of Architecture entirely of Hewn white

Colour'd Stone in the shape of a Cross the upper and Western[99] end of which makes the Choir. The Cupola is design'd to be painted but is not yet begun. The inside is very well finish'd and looks extreamly noble and indeed I have thought it is impossible to enter its lofty and proud isles without haveing ones Ideas somewhat exalted to that being which dwells peculiarly there; at least the Cupola has in its distance its figure and extensiveness from the Eye so near a resemblance to the vast Arch of Heaven that one cannot look up to it without a sensible reflection on its Almighty Inhabitant. How far the Painting of this Cupola may help or may destroy these reflections I cannot tell, but I am sure it may be dispos'd to vast advantage in this way if the painter thinks it proper. When I have related the Ideas of Veneration and Pleasure I felt on entering St Paull's, I may with the greater Freedom confess those things that indeed shock'd me in this Fabrick, and first That [**35**] which indeed is rather an unhappiness than a Fault, it stands in the most tradeing opulent part of the City where ground is wonderfully precious & therefore is choak'd up with houses & never can have a sufficient Area to do justice to its prospect however as well as I could put it together from so near Views as you are forc'd to take its outside had indeed something prodigiously Magnificent and surprizeing yet I could not but fancy the number of Pillars, Carveings and Ornaments, the several breaks and different Figures of the building appear'd very Gothick and unnatural and I am sure the whole had pleas'd me much better if it had wanted these Superfluitys, you see I am not Architect enough to attack St Pauls in form but however if I have not the Art I have not the Envy of a Critick and you are sure to have Nature's judgement unbyass'd in what I say; in this view I shall venture further to confess that after haveing walk'd with great Pleasure thro' the magnificent body of Pauls & haveing there rais'd my Ideas to something near divine when I enter'd into the Choir it fell surprizeingly short of my Expectations and I found it no ways answerable to the Grandeur of its avenue; 'tis little and mighty dark, the Altar-Piece extreamly mean and Comon, no painting or gilding scarce anywhere appears, and that I might not be quite put out of Conceit with Pauls I resolv'd to beleive that this is to be alter'd and made infinitely finer. A merry [**36**] man in Our Company say'd pleasantly enough on entring the Choir; he thought he underwent Jeremiah the Prophett's Fate who was thrown out of a Prince's Palace into a dirty Well. Before we leave this Church I must not forget to mention its vast Vaults, which extend I think almost under the whole and are very large and Commodious and possibly in future Ages may make a Labyrinth as famous as Diana's at Ephesus. Very distant Ages if the world lasts long enough may possibly make this Diana's Labyrinth as well as that, for you must know where this glorious Fabrick stands was once a Temple of that Goddess, so that 'tis with more Justice than is generally known that the Apostle of the Gentiles

99. It would appear a mistake for the Eastern end.

here triumphs over Idolatry in a Temple dedicated to his name. The Iron rails round the Church yard are also remarkeable, they are mighty strong massy work, were turn'd all in an Engine with which they turn Cannon and cost as Mr. Wren told me ten thousand pounds at least.[100] In the yard before the great door stands a fine Marble Statue of the Queen, which is not ill done, by one Mr. Bird who perform'd most of the fine carveing about the Church. The Prints and Platform of the Church which you have commonly sold about are very just and well done tho indeed they do by no means give one a notion magnificent enough of its Structure; I think I could have rece'd a much better Idea of it from a **[37]** Model of the Church which they take Care to shew in Oak about 10, or 12 feet long, which is very pretty and exact in one of the Chambers above Stairs over the Isles, this is not indeed the modell of the present building but of what was once design'd to be executed and I think it at least as beautifull as the present, they shew you there a second Modell somewhat broken which represents the Figure of Pauls as it stood in its ancient Gothick Shape; and on the opposite side of the same Isle you have a Correspondent Chamber which is already as both are design'd to be made into a Library. The Rooms are Capacious enough but somewhat dark for that purpose. From hence they carry you to the Cupola, which is as indeed I must observe to you, the whole building appear'd to be in a surprizeing manner perfectly well executed so that the nicest Eye cannot discover the least Errour, And that here the Ear as well as the Eye bears witness of the exquisiteness of the workmanship, there are under the Cupola in the Gallery above certain whispering places Where the lowest voyce shall by the round Figure be propogated so as to [be] Audible to those that are diametrically opposite, tho at the distance at least of a 100 feet or more, nay I do affirm to you I try'd it with my Strikeing watch and could distinctly hear the hours and Quarters struck and was assur'd in the Silence of the night the very Ticks of its going are perceiveable, from hence you may go to the Top of the **[38]** Cupola and so if you please into the Golden Ball which stands above it 365 feet high, which tho' it looks no bigger than a gilded Pill below will hold at least 12 or 15 men, you have there a very good prospect of the City of London, the Thames and many of the publick places in the Town, and tho the fog and smoak did much hinder the Prospect yet I am confident I saw a surfeit of houses and [at] least Six times as big as Dublin from thence. There is nothing perhaps more remarkeable in all this mighty Fabrick than that a building such prodigious Extent at least 600 feet long and every way proportionable of such expence as to have cost by a duty on Coals imported from time to time into London for these 50 years past above half a million should have been begun and now almost finish'd by one and the same Architect Sir Christopher Wren.

I know no occasion so proper to mention, the great old Gothick Abby

100. Actually £11,700 was spent on the ironwork.

of Westminster to you as this is, and tho' Pauls has too much Gothick and this too much Magnificence to make them perfect exact Standards of their different manner of buildings, yet I do believe in London one cannot have a juster taste of the difference of these Manners than by compareing with Care these two mighty buildings. The contemplation of this Church and [**39**] its Gothick Magnificence put me in mind of a Reflection I have often made in reading the Schoolmen[101] which I thought applicable to these Gothick Architects, also that they both shew a very great and noble genius very ill-managed, and as I doubt not but the Schoolmen instructed in modern Philosophy had made great Figures in the world, so I am confident our forefathers that could build Westminster Abby and Westminster Hall if they had had a right Taste had made stupendious Architects. This Abby beside that it is very large, within ten feet in length of St Pauls, and loaded with infinite Ornament of the Gothick manner is remarkeable for being very ancient haveing been founded by Edward the Confessor who granted them their Charter and will'd this place to be the place for the Coronation of the Kings of England and accordingly it continues and I am told they preserve here several Regalia and Jewells given them by Edward the Confessor but I did not see any of them, You here see ten or 12 little Chappells now used as burying places for the nobility and in them many noble Monuments of Marble to their Memorys, in one I observ'd two unbury'd Coffins in which it seems lye deposited two Foreign Ambassadors who dying here wait there [**40**] to be sent for home and by the way have waited there these 20 years, one of these Chappells the finest and largest is called Henry the 7^{ths} Chappell built by him for the Royal Family and in it lay several of Our Kings. In this apartment of this Abby meets the Convocation of the Clergy of England when they are call'd together by the permission of the Dean and Chapter of Westminster; to do the Royal Duty great Honour, and that those who prais'd them liveing may be at hand to support their trembling Patrons on a more dreadful day there bye spread around this abby thick Crowds of Poets and Ingenious Men; Heroes and Parasites have here their Monuments, Virtue and Power and vice and Wit and Honour are here remembr'd; and if the wand'ring Manes should come one night to visit each its property I do beleive it would make the most mix'd Company that ever was seen. If there would be a strife for the Crown among the Kings the Poetts would have the Struggle for the Laurell and in all the Rout perhaps not one but might be reasonably content to lay down again in his grave and be just as he was. Forgive these Whims and let rather put you now in mind of one of the noblest Largest single Rooms that is perhaps in all the World it Joyns to the Abby and is the famous Westminster Hall where many a cause has pass'd the dreadful Tryal of the most August Assembly in this little World, the British Parliament; [**41**] Here all impeachments are heard and all great Assemblys and Coronations or any

101. Most likely meaning medieval writers and scholars.

like solemn State occasions are held, it is 230 feet long 70 broad and 90 high, and is computed to contain at least 14000 persons without Galleries built as 'tis say'd by William Rufus. The Roof is of Irish oak & a very fine piece of Architecture haveing no Columns to support that vast breadth, which makes it appear prodigiously lofty and surprizeing when one first enters; in different parts of this magnificent Room are the Chancery, the Queens bench and Comon-pleas-Court in term times; all along the Walls several different Shops of Books, Prints, Toys and other things for Sale; but among all the Glorys of it the most distinguishably Ornamental ~~to all but British eye~~ are above 160 Standards and tatter'd Colours which long had led the French in their unbounded Conquests 'till Blenheim and the Duke of Marlborough sent them (a well imagined disgrace) to adorn the great Palace of Justice in the British nation, and here they hang on both sides high in a range fix'd in the Wall as yet out of the reach of little Fellows, on one of them, a white one, I observ'd a pretty poeticall Fiction for a Motto and it was this Victoria pinget.[102] But now what Faith shall Implore if any Stranger chance to read these papers shall know that thro' this very Hall thus adorn'd is the way to the Parliament houses of Great Brittain. **[42]** Who will believe that Gratitude can be so much a Stranger to Our Hearts that Spoyls like these can unregarded hang before a nation's face at home where he that won 'em for these very Spoyls now reigns a Prince but yet a banish'd man abroad. Yet so it is thro' Westminster Hall you go to the House of Comons and the House of Lords which join to it and are handsome enough but have nothing remarkeable except it be that in these houses meet an Assembly which united might give Laws to the World, divided as they are cannot govern themselves; in these reflections I was glad to turn my thoughts to the Antiquity of the building which it seems was Edward the third's Palace and here therefore they shew you the room that was his Bedchamber, now call'd the painted Chamber where the Lords and Comons hold Conferences; here you see the Cellars in which in the Popish plott the Magazines of Powder were lay'd and the door thro' which they were brought in now stop'd up with a Wall. The Destruction of the Spanish Armada is also well done and remarkeable in some fine hangings in the House of Lords presented to Queen Elizabeth by the States of Holland. And here I should not do Justice to the Belles Lettres and Politeness nor to my own Inclinations if I neglected mentioning with the greatest respect and Esteem their Patron my Lord Hallifax. He lives in one part of these **[43]** old Westminster Buildings and has made it a very convenient and very handsome house; he has three noble large rooms in one apartment perfectly well furnished with some original Pictures and many copies of the best hands, which are very well worth seeing, but that which deserves a more distinct regard is his Cabinet in which are several true ancient Roman and

102. French banner of the Fifth Dragon Regiment with the motto '*Victoria Pinget*', meaning 'he becomes more beautiful with victory'.

Grecian Bustes, some of Porphyry and some of Marble, I believe at least 20 or 30, particularly a Fine Antinous, a Tully, a Venus, A Caius Marius with several very good Copys of the best antient Statues, among these in this Gallery are his books which are a very valueable Collection of the most Curious, distinguish'd into Classes according to the Sciences and the Shelves so contriv'd that every two contiguous apartments are joyn'd at the forepart, in the middle by two Hinges so that you can shut 'em up and carry away the Books without disturbing any one which is very convenient especially in case of Fire, Over the Shelves are hung several original Portraits of Ingenious Men of England, principally Poets, so that in the whole and in his manner of shewing them, which to me has been allways extreamly obligeing I think I never met any thing yet so completely satisfactory in its kind as this. If my Lord Hallifax be extreamly polite in his Studys he is no less so in his other parts of life. Elegance [end of folio **43**]

[folios **44–57** missing]

[**58**] it deserves to be seen on its own, 'tis peculiarly neat and elegant in all respects as indeed the whole house is, we saw here another valueable collection of Manuscripts, Journalls of both houses of Parliament for a long space of time which are very great Curiosities. I remember it was M[r.] Dale carri'd us to this place whom I just now mention'd, I must not forget to tell you this Gentleman is one of the greatest instances I have known of natural Genius, he was brought up without any manner of Learned Education a Cloathier and continues so still and actually now keeps his Shop near the Exchange and yet in the Intervalls of busyness has made so good advancement in Learning that he is now a very good Scholar understands French, Italian, Latin & Greek and has a pretty good taste of most kinds of knowledge whatever and indeed shews an uncommon Genius in his Conversation which I was much pleas'd with.

I believe by this time I have tir'd you with Philosophy, we will take a Respite & go down the River to the Tower where are Curiositys of another kind, but first let me observe the great Surprize I had in seeing the thousand little boats and lighters that ply up and down on this great River, which as it runs parallel with the Town has given great opportunitys to this manner of Carriage and maintains as I have [**59**] been well assur'd by this means at least 1500 or more seafaring men, which are at least fitter than others for manning out the Royal Navy, These little boats row with very great swiftness & are indeed extreamly convenient both in going up and down the River, and passing it over; in one of these I went down to see the Tower; it will be needless to tell you that this is a kind of Ancient Gothick Citadell which governs the Citty and the River and that there is here a Prison which is the State Prison and Prison of the Parliament; I shall only tell you what I observ'd here particular and remarkeable and first, from the Water they shew'd us an old Wall, part of the Tower, say'd to be built by Julius Caesar, however certainly Antiquarys can evince these marks of Roman

Conquests, in Brittain more modern Spectators are sure they see in the Tower more sensible marks of Brittains Conquests and Glory, and I was not half so sure that Julius was a Conqueror here by seeing his Wall as that the Duke of Marlborough was a glorious one abroad by several pieces of Artillery, Standards and Colours taken under his happy Conduct of this War from the French and lay'd up in the Arsenal which is the first thing they shew you in the Tower; it consists of two Storys, on the Ground Floor which is about 200 yards long you have the heavy Artillery almost all of brass to the number I think at least of two hundread Pieces with all kinds of **[60]** appurtenances necessary for them, among these you see severall Curiositys, as several French peices dischargeing three, and one I think seven Balls out of as many Bores at once; Kettle drums and other Spoyls taken at Hockstel,[103] a Mortar that shoots nine, and another thirty Shells at once, little hand Mortars for Grenadoes, Pontoons, A smith's forge to work in marching and many such military Contrivances, but when you go up Stairs you have on the Second Floor as large a Room and in it a most beautifully dispos'd Armory for above 70 thousand men and I was assur'd when their Arms that are now abroad in the Service shall be return'd they will have arms for above 150 thousand. This Room is very lofty and wainscotted to the Top, but so beautifully are the Walls cover'd with Swords, Pistolls, Carbines and Bayonetts, dispos'd into such odd figures so exactly and kept so close with flatt Pillars made of Pikes set close to one another by which the Pannells are divided with Pistolls for their Chapiters, that nothing can be imagin'd more curious; in the Area of the Room are Racks up to the very top for firelocks, supported by Pillars of Pikes at either end of these. I believe there may be 20 or 30 or perhaps more in the whole length so that the Walls and the Area of the room is entirely fill'd with arms insomuch that you have but just the Space of 8 or 9 feet to go between these Racks, and the Wall all round which unhappy Straitness and that the room is a **[61]** little dark are the only things I observ 'd in which I could not esteem it as magnificent a Project in its kind as I could conceive; It was impossible however for a man of good nature & benevolence to mankind to walk with Pleasure in this Repository of Death & Destruction, and however I might approve of the Prudence I could not but grieve at ye necessity of such Provisions. What Violence, what Wickedness, what Rapine in our Neighbours does such a mighty apparatus presuppose, and who that reflects on the Happyness, as well as the Virtues of Peace and Tranquility can see such furniture of War, of Pain & Torture to his Fellow Creatures, in effect, I always have abhorr'd the Trojan Horse and its Bowells of Death and I could not but see this mighty Arsenal with the same Impressions of Horror & left it with like Reflections of Regret and Concern; From hence we were carry'd to another much more peaceable and harmless Armory, and indeed I think very near as

103. Höchstädt, a town 4 miles south-west of Blenheim.

Ornamental, this they call their horse Armory, it consists entirely of horse Furniture formerly us'd and Suits of Armor of which I think they reckon about 200, many of which they shew you formerly us'd by several of Our Kings, some of which are neatly inlay'd and some gilt, they show'd us one compleat Suit which was above 7 feet long formerly as they call'd it John of Gaunts Duke of Lancasters [**62**] from whom the present Duke of Beaufort is descended; it is certain he is related to be monstrously large, it is from some Exploits of his that the Portcullis had place in the Arms of Great Britain. they shew you here also an old Sword said to be the very one taken from a French General by Conreye Lord Kinsale for which that Lord's Posterity may be cover'd in the King's Presence. From hence we went to a third Armory which consists principally of old Suits of Armor and other old Arms taken from the Spanish Armada in Queen Elizabeth's Reign, they shew you here a wooden Canon, a Rack, the Axe by which the Earle of Essex lost his head[104] and some other old Curiositys which I have forgotten. The Customs of the World has made these two last Armorys now useless in the military Art I could not but wish the Virtue of the World might make all three so; Haveing viewed the Defence we went to see the Ornaments of the Crown of Great Britain which are here a great part of the Show at the Tower; they carry you with great Solemnity into one of the little old Towers of the building thro'[105] a dark passage into a small room which is divided by an Iron Grate; on one side you stand and on the other appears a Table Spread and two Candles to expose these precious Regalia; at length appears a very venerable person to whom the Care [**63**] of them is given, who seems to have no better recommendation to this Office than that he looks to despise the trifles; having pay'd your money in great Order he shews this Scepter and that Diadem, this Jewell and t'other Crown and all the while I am assur'd they have not the true Crowns nor Jewells here at all, and I remember very well a Gentleman told me of good Credit he had this from my Uncle Lucas who you know was Governour of the Tower; we shall not dispute them their gainful Shew, all I can say is some may be counterfeit and some real or all the former for ought I can see and no body the wiser in the manner they are shewn; for my part I am sure if they be real they are of immense Value and should not be trusted as they are, for you must know with all this Rout and Iron Grates, there is put 2 Slittdale doors between them and the open yard and I have often wonder'd since they make such a work with them some hardy Fool has not been made believe they were real and attempted to carry them off. I remember to have heard of a King that us'd to say in Mirth that Man could do any thing that had his Sword in his hand, a Crown on his head and money in his Pocketts; We have arm'd and crown'd Brittain and before we leave the Tower we must give her the third [end of folio **63**]

104. No longer in the collection; perhaps its existence was a fable recounted by his guide.
105. The Martin Tower.

[folios **64–76** missing]

[**77**] With [torn page] the external variating Air; [torn page] as inexplicable yet that is certainly the [torn page] I have told you, considering the greater influence [torn page] must have on these glasses than on common ones, that their Degrees of riseing and falling must by the Inequal or at best uncertain contents of different parts of the Tube be very hard to adjust truly and cannot be done but by the help of a common glass. I do by no means esteem them unless they are made with better chosen pipes of true Cilindrical Diameters and more exquisitely made than usual, however being fram'd but with care tho' not for comon use. I think them very proper for Philosophical observations of the Change of the Air. After this I think we visited Mr. Hawksbee, Operator to the Royal Society, who shew'd us several Curious Air Pumps of his own makeing and among a painted pastboard of the annex'd figure which by wheeling round gave a sensible Demonstration of what Sir Isaac Newton advances of the Composition of white in his Book of Colours (Fig. 10). From him we went to one Willsons, a glass grinder, we saw his Microscopes of Single glasses which are indeed excellently good and worth any bodys enquireing for; I think he told me he sells a compleat case of them for about 3^{li} which is not by any means [**78**] dear; From hence I went with my friend Signore Bianchini the Italian to see one of the greatest Curositys of its sort by much the greatest now in England, and this is a Collection of Antiquitys got together at vast Expence by one Mr. Kempt, This person was formerly a sword-Cutler but being naturally an ingenious man, and haveing a head turn'd to Medalls and the like has taken up this way of life, and collected great quantitys of Medalls and other Antiquitys which he sells again as I believe at extravagant rates and lives by it. The Connoisseurs assure you he can counterfeit Medalls & does actually better than any man in the world & I remember Signore Bianchini suspected grievously most of his rare ones. However he has certainly with these several very great Curositys that are truly Antiques, the Principal of which I shall tell you; He has a very beautiful Bass-Relievo of the three graces of about 9 Inches length, another larger of Mars & Venus & Cupid, A Socrates of the same, a Senator of Rome in his Robes with the Ancient manner of Books represented by him, another representing a Colony probably in the East Indies, there being an Indian Bull and an Indian tree represented plainly there in which is possibly a great Curiosity, a head of Cupid or some young Boy admirably well done of two Colours in the manner of an Aggat Gem, all the Bass relieves are of marble, He has great Collection of the Lares and little Statues of the antient Gods [**79**] severall of their Lamps & Vessells, particularly one very fine brazen Lamp formerly Mr. Spons and describ'd in his Book an Ancient Lead Urn and of Statues about 3 feet high I think he has got 8 or 10 pretty good ones, and about twice as many Busts among which one of Venus and one of Antinous are very good. He shews

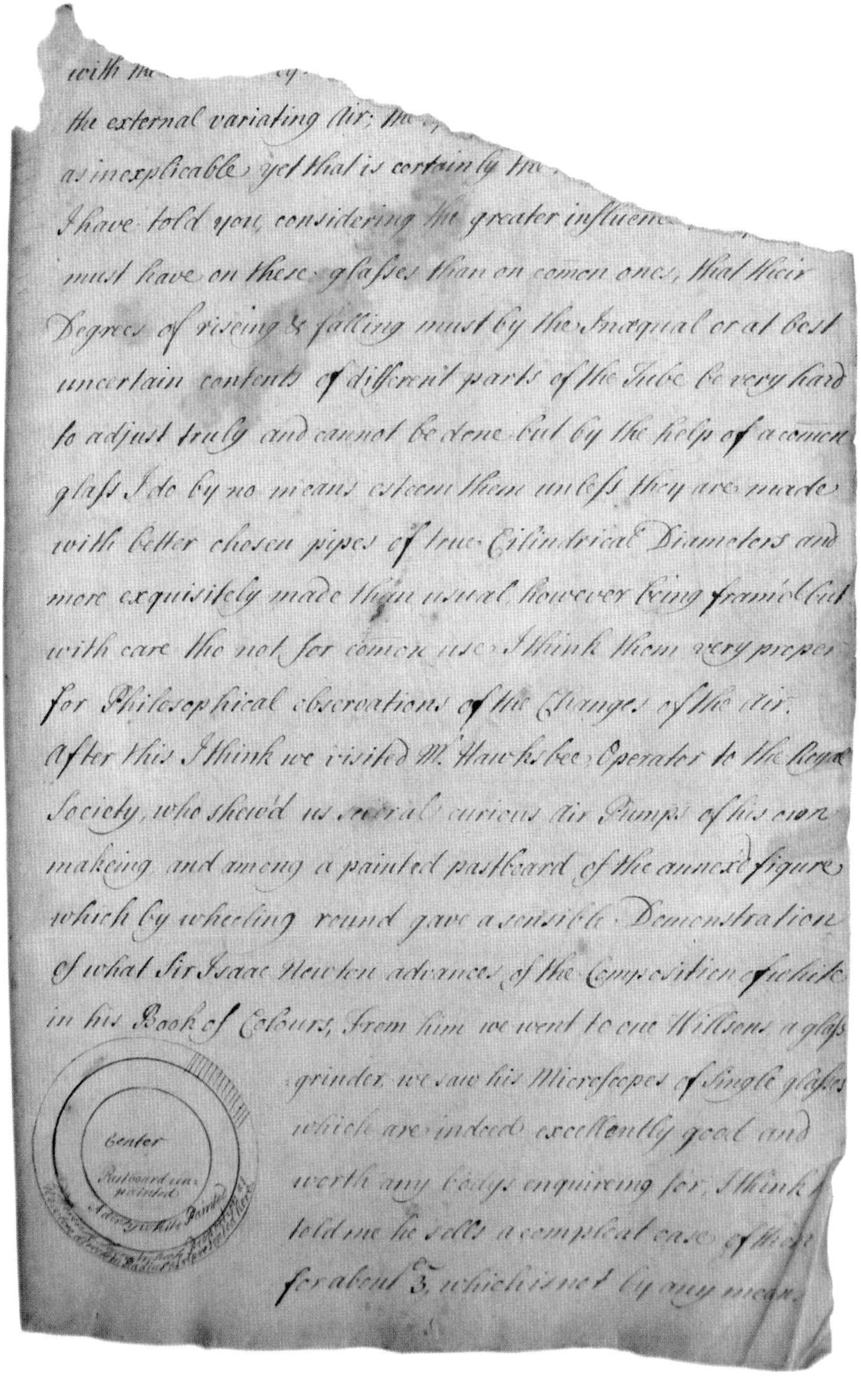

with m... ...y.
the external variating Air; the...
as inexplicable yet that is certainly the...
I have told you, considering the greater influence...
must have on these glasses than on comon ones, that their
Degrees of riseing & falling must by the Inequal or at best
uncertain contents of different parts of the Tube be very hard
to adjust truly and cannot be done but by the help of a comon
glass I do by no means esteem them unless they are made
with better chosen pipes of true Cilindrical Diameters and
more exquisitely made than usual however being fram'd but
with care tho not for comon use I think them very proper
for Philosophical observations of the Changes of the Air.
After this I think we visited Mr Hawksbee Operator to the Royal
Society, who shew'd us several curious Air Pumps of his own
makeing, and among a painted pastboard of the annexed figure
which by wheeling round gave a sensible Demonstration
of what Sir Isaac Newton advances of the Composition of white
in his Book of Colours, From him we went to one Willsons a glass
grinder we saw his Microscopes of Single glasses
which are indeed excellently good and
worth any bodys enquireing for, I think he
told me he sells a compleat case of these
for about £3, which is not by any means

Fig. 10. Molyneux manuscript showing pasteboard illustration, f. 77.

you two very perfect Mummys with some colour'd Hierogliphicks on one, still in their ancient Coffins and several little brazen Images of Osis & Osiris, & other Egyptian Deities which he found in the Body of a Mummy he open'd, two Sphinxes of Brass, a helmet of the same work & antiquity with Dr. Woodwards Shield which I am sure you have seen the figure of, this he does not esteem very ancient, among many antient Marbles with Inscriptions he has one pretty large one which Signore Bianchini told me was very valueable, 'tis an imperfect Fragment of a List of a Roman Legion and their quarters. The last curiosity I shall mention here is an old Bak'd Earth Lamp found near Paul's Church London on which tho' much defac'd you plainly see the figure of an antient Temple and a River with a Boat on it running hard by; this possibly may be the Representation of Diana's Temple which formerly stood there and to which this Lamp possibly belong'd. When I have mention'd Antiquitys I am sure you will expect I should give you an Account of my Lord Pembrokes Curiositys, in that way which indeed I have seen with as great Pleasure as my [**80**] Lord shews them and this I assure you he is so good natu'rd & loves so much to instruct a Stranger he does with the utmost willingness and Satisfaction. His Medalls in Silver & Gold & Brass are very numerous and perfectly good ones of each Sort and well preserv'd, He wants very few of the most rare, and is esteem'd to have the most Compleat Collection of every Series that can be seen in England, besides a few very rare ones of the Roman and Greek Medalls, which he desir'd me to observe; he shew'd me some Arabick Medalls of Mahometan Princes & some Phoenician which he esteem'd great Curiositys as not being often to be met with in any Cabinets, A much more valueable one I think he preserves and values it as it ought to be among his ancient weights which are all very curious and well preserv'd, and this is a Boss of Brass weighing five Asses; it was found in the Catacombs at Rome, is very little worn and decay'd and is thought to be the only one that is now known in the World, it is of an oblong Square figure of about six and four Inches with a very high Relieve on one side of an Ox the other plain. Pliny gives the Description of this Weight but confesses that even in his time they could not meet with them and that he had never seen any one of them himself, My Lord shew'd me also another very great Curiosity which is a large Folio containing a Prodigious Collection of old and new Prints of all the different Masters & Inventions in Etching & graveing [**81**] on all kinds of Plates from the first beginning of that Art in the year [gap] by Bogliggliolo to the present time, among which I thought two very remarkeable, one done by Melan representing the Head of Our Saviour and perfectly plainly too of ye natural bigness and yet the graveing of the Plate was in this manner: they begun at the Tip of the nose and made from thence a spiral line whose distances were inconceivably small & exquisitely equal and this very Line was thus continued to the very outskirts of the Print, and yet without any kind of

Lines or graveing whatever, a cross was so proportioned to be deeper & Stronger Lighter & weaker in different parts, as to make the Shadow and Lineaments of the Face, the Curling of the Hair, and the Crown of Thorns and every part very distinctly & plainly appear, the other Print I remark'd was made by one Teillier, in different colours, I think there were three or four distinctly different, which is an Art of great use in the History of Nature but is now entirely lost. He shew'd me also a very valueable collection of all the Pretious Stones & best Egyptian Marbles that are known in the World neatly pasted into [gap] which make the leaves of two volumes in folio, and are neatly bound up like other books, these besides a great many curious Prints especially many of La Trevit and a very neat collection of Books principally the Classicks, above 60 of them of Elzevir Printing and several Editions older than Fabritius mentions **[82]** several Curious Books of Travells and Antiquitys do make up my Lord Pembrokes Study. I must not neglect to tell you that in going to it you pass thro' a Gallery where are several valueable Pictures so that those that are not curious in Antiquity will be very well pleas'd however to visit his Cabinet if it were but to see its Avenue and to observe the great Courteousness & Pleasure he shews in receiving such a visit.

I am Y[rs] Sincerely

LETTER 3

(complete: ff. **83–97**), 14 February 1713

SCA D/M 1/3

As would be expected from an early eighteenth-century gentleman's library, Molyneux's books reflected the pursuit of 'enlightened' learning which, in part, mirrored the owner's identity and gentility. Amongst his collection were volumes on astronomy, history, politics, religion and the ever-burgeoning genre of antiquarian-based county history. Moreover, typical of the mindset of a natural philosopher, books on nature, particularly gardens, geology, horticulture, landscape and topography, also featured strongly.[106] His next letter, dated 14 February '1712/13' and written from London, provides his reader with unexpected glimpses of contemporary fashions of the English landscape garden through interpreting the positive appeal of the wildness of nature over the displeasure of restrained contrivance.[107] As such he shares Joseph Addison's (1672–1719) ideological beliefs portrayed in his 'Pleasures of the Imagination' essays published in *The Spectator*.

In his search for an 'enlightened' authenticity, Molyneux visits Hampton Court, Bushy Park, New Park (the home of Lord Rochester on the River Thames near Twickenham), Fulham Palace (the residence of the Bishop of London), the Chelsea Physic Garden and Kensington Palace. Only the two royal palaces carry some description of their interiors and contents. His interest in matters of horticulture also prompts a visit to see the pre-eminent nursery garden at Brompton Park, created by George London (d. 1714) and Henry Wise (1653–1738). London and Wise had, according to Addison, 'a fine genius for gardening', and through the pages of *The Spectator* he lauded them as 'our heroick poets'. Molyneux's journeys outside the city, with one notable exception, appear to have been taken alone and, once back in London, his focus returns to buildings and collections with accounts of St James's Palace, Somerset House, the Banqueting House in Whitehall, Buckingham House, Marlborough House and Berkeley House. Before the letter ends, he presents in some detail his meeting with the famed antiquarian and collector Dr John Woodward, Professor of Physic at Gresham College in London. The letter is complete.

106. *A catalogue of the library of Samuel Molyneux deceas'd ... consisting of many valuable and rare books ... with several curious manuscripts* (London, 1730).
107. Refer to 'Manuscript' section for clarification of the dating system.

Of the Tudor Hampton Court Palace (Fig. 11), Molyneux describes the ancient buildings simply as 'a very fine old Palace built by Cardinal Woolsey' (f. **83**). Ignoring the old parts of the palace, his narrative shifts unequivocally towards the modern and to Sir Christopher Wren's unfinished interiors. From the Clock Court colonnade, Molyneux's tour starts with the King's Staircase where he comments briefly on Antonio Verrio's murals of *c.* 1700 based on Julian the Apostate's satire *The Caesars*. Curiously, he does not comment on their problematic iconography, nor does he point out the spectacular wrought-iron staircase balustrade by Jean Tijou. Next he visits the 'well dispos'd Armory' (f. **83**), said in the earliest guidebook to contain 'Arms for five thousand Men, artfully dispos'd in various forms'.[108] This 'Guard Room' or King's Guard Chamber was designed in *c.* 1700 by Mr Harris, gunsmith to King William III, under the direction of Wren and John Talman. No specific mention (perhaps in view of their unfinished or unfurnished state) is made of the King's State Apartments on the first floor, which incorporated the first and second Presence Chambers, the King's Audience Chamber and his Drawing Room, although Molyneux does mention 'a plein pied furnish'd with Tapistry' (f. **84**) which corresponds with the 'open walk' or long corridors hung with decorative tapestries as described in the 1742 guide. Molyneux confesses his pleasure on entering the Queen's Gallery, situated on the eastern side of the palace overlooking Fountain Court, where he inspects the nine canvases known as *The Triumphs of Caesar*. Painted between *c.* 1485 and 1494 by Andrea Mantegna (1430–1506) and commissioned to inflate the authority of the Duke of Mantua, who by implication inherited the imperial mantle of Caesar, the canvases were purchased by King Charles I and were hung at Hampton Court by 1631. Although valued at £1,000, for a potential sale after the king's execution, it was considered too risky to move them from their position and consequently, between 1690 and 1702, they underwent a programme of restoration by Parry Walton and Louis Laguerre.[109] According to the 1742 guidebook these paintings were displayed in the Queen's Drawing Room, adjoining the Gallery, in order to conceal the 'badly executed' wall paintings by Verrio.[110]

Completed by Wren in 1699, the Cartoon Gallery held yet more remnants of the late king's collection, seven of the ten cartoons depicting the Acts of the Apostles, which were painted by Raphael between 1515 and 1516 as tapestry designs for the Vatican Palace. These cartoons, the largest of Raphael's works to be found outside Italy, meet with some ambivalence

108. George Bickham, *Deliciae Britannicae; or, the Curiosities of Hampton-Court and Windsor-Castle* (London, 1742), p. 30.
109. Oliver Millar (ed.), 'The Inventories and Valuation of the King's Goods, 1649–1651', *Walpole Society*, 43 (1970–2), p. 186; Christopher Lloyd, *Andrea Mantegna: The Triumphs of Caesar* (London, 1991), p. 20.
110. Bickham, *Hampton-Court*, pp. 85–6.

Fig. 11. Hampton Court from the south showing Wren's additions to the palace (1689–94) and Franco-Dutch formal garden, *c.* 1702.

from Molyneux, adding fuel to the contention that his intellectual response is influenced less by the art itself than by contemporary opinion. Instead of an inspired account of the artwork, he digresses into a learned familiarity with contemporary source material, in this case 'Dr: Clarkes new Edition of Caesars Comentarys' (London, 1712) (f. **84**). Samuel Clarke (1675–1729), Rector of St James's in London, was a noted natural philosopher who in the same year published a celebrated, yet controversial, treatise, *The Scripture Doctrine of the Trinity* (London, 1712), a copy of which was catalogued as part of the Pitt Collection.

Molyneux also mentions the prints of the cartoons '[that were] handed about every where' (f. **85**). The first set of engravings by Simon Gribelin (1661–1733) was published in 1707 but others followed, most notably a set published in 1709 by Henry Overton and a set of mezzotints (mentioned on f. **85**) by John Simon and published by Edward Cooper in 1710.[111] During his visit, Molyneux encounters the French engraver Nicolas Dorigny (1658–1746) at work on a new set of engravings. Dorigny, who first came to England in 1711, earned his reputation through his portrayal of these cartoons. Amongst his champions was *The Spectator*, which wrote

> The whole Work is an Exercise of the highest Piety in the Painter; and all the Touches of a religious Mind are expressed in a Manner much more forcible than can possibly be performed by the most moving Eloquence. These invaluable Pieces are very justly in the Hands of the greatest and most pious Sovereign in the World; and cannot be the frequent Object of everyone at their own Leisure: But as an Engraver is to the Painter what a Printer is to an Author, it is worthy Her Majesty's Name, that she has encouraged that Noble Artist, Monsieur *Dorigny,* to publish these Works of *Raphael.* We have of this Gentleman a Piece of the Transfiguration, which, I think, is held a Work second to none in the World.[112]

George Vertue later added that

> [...] several gentlemen of note travelling to Rome there found Mr. Dorigny who was then in the highest reputation for several engraved works after Raphael. These got him justly the reputation of the first engraver in Europe for which reason several Curious persons persuaded & engaged him to come to England to undertake those Famous Cartons at Hampton Court painted by Raphael [...] From his coming to England I may justly date the rise of the reputation of the engraving.[113]

111. Chia-Chuan Hsieh, 'Publishing the Raphael Cartoons', *The Historical Journal*, 52, no. 4 (2009), pp. 899–920.
112. *The Spectator*, 19 November 1711.
113. Quoted in Nicolas Dorigny, *The School of Raphael: The Student's Guide to Expression in Historical Painting*, ed. Tom Richardson (2010).

Dorigny's commission commenced in spring 1712 and took seven years to complete, and the finished engravings were presented to King George I in April 1719. It appears that Molyneux was already familiar with Dorigny's work as one of his books also survives in the Pitt Collection.[114]

It is a recurring theme in Molyneux's narrative to reject the Baroque style in favour of the classical revival. Little analysis is made of the architectural merits of the Baroque Hampton Court Palace and, when he moves into the gardens, Molyneux expresses himself disappointed. Baroque gardens represented huge investments and Hampton Court had one of the most lavish in Europe but, like many of his contemporaries, Molyneux feels that its French geometric design (derived from the Palace of Versailles) is unsuitable for English tastes. Instead of a detailed description of Hampton Court's 'Walks, Labyrinths & Situation' (f. **86**) he refers his reader again to the 'Comon prints', in this case most likely Johannes Kip's bird's-eye plan (*c.* 1705), illustrating the scene from the south-west.

Moving on from Hampton Court, the private recreational garden at Bushy Park, created by its Ranger, Lord Halifax, once again offends Molyneux's sensibilities. He considers only the cascade as being of note; this feature was installed in *c.* 1710 when Matthew Banks was paid '[...] for making Bayes to keep the Water back while the Cascade was putte down in Bushey Parke'.[115] John Loveday added more details on the water features when he wrote, in 1733, of 'Lord Halifax's brick house in Bushey Park and his "neat box" near a cascade which in the front presents you with a Rock of Stone and burnt-Iron, on each side Canvass painted to represent the perspective of Caves — with success', adding in 1736, 'The two painted Caves on Canvass have these words painted on them — *'Ts, Carwitham Fecit 1736*. They are done inimitably well to deceive'.[116]

Molyneux's disenchantment shifts with a jolt when visiting New Park in Surrey. Situated alongside Ham House and overlooking the River Thames towards Marble Hill, New Park was one of the most remarkable gardens of its day, having extensive formal walks, parterres, pools, terraces and an elaborate wilderness garden. Yet, despite such a magnificent setting, the house of Laurence Hyde, the 1st Earl of Rochester, was extremely modest. At the highest part of the garden was a mount which remains the only surviving fragment of this once illustrious landscape now absorbed into Richmond Park. New Park accords with Molyneux's idea of 'perfect Satisfaction'; he adds, 'I think I have never yet seen any piece of Gardening

114. SCA, *Pitt Collection*, p. 38, no. 55.

115. John Harris, 'Water Glittered Everywhere', *Country Life* (6 January 2000), p. 46. Matthew Banks may well have been the son of Sir Christopher Wren's master carpenter, Matthew Banckes.

116. Markham, *Loveday*, pp. 149–50 and 247.

that has so much as this the true taste of beauty' (f. **87**). Folios **85** to **90** have previously been published in *The Genius of Place: the English landscape garden 1620–1820*,[117] in which Molyneux's description of the grounds at New Park (or Petersham Lodge, as it is referred) is described as 'possibly the earliest account of an actual English landscape garden, as opposed to merely theoretical formulations of the idea', adding 'That Molyneux appreciated the radical nature of the gardens there is clear from the contrasts he makes with others in the French taste of "regular manner of greens and gravel gardening"'. His progressive approach to horticulture and garden planning was well represented in his own library and may also, in part, have been influenced by articles published in *The Tatler* and *The Spectator* and by the anti-Baroque polemic *Letter Concerning the Art, or Science of Design* by Anthony Ashley Cooper, the 3rd Earl of Shaftesbury (1671–1713), which may well have been in circulation in manuscript form by October 1712.[118]

Curiously, Molyneux does not discuss in any detail his visit to two of the finest exotic gardens in London at the time — the Chelsea Physic Garden (Fig. 12) established by the Society of Apothecaries and the fine horticultural collections of the zealous Protestant Henry Compton (1632–1713), Bishop of London, that were described in 1713 as 'a greater variety of curious exotic plants and trees than at that time had collected in any garden in England'.[119] Both gardens were started within two years of each other, Chelsea in 1673 (where the first cedar trees in England were planted) and Fulham Palace (now a museum) in 1675 after Compton was suspended from office by the Catholic King James II, and reflect the contemporary popularity of collecting foreign specimens. Nor does Molyneux dwell unnecessarily on the merits of the Brompton Park nursery (100 acres on the site of today's Kensington museums) of which London and Wise had been in sole partnership of since 1689. Such omissions reflect his bias towards garden design over that of a plantsman, as Molyneux himself confesses, 'I have not Botanicks enough to tell you all the Exotick Plants' (f. **86**).

In September 1712, shortly before Molyneux's visit, *The Spectator* gave a description of the garden at Kensington Palace (Fig. 13):

> If as a critick I may single out any passage of their works to commend I shall take notice of the portion of the upper garden at Kensington which at first was nothing more than a gravel pit. It must have been a fine genius for gardening that could have thought of forming such an unsightly hollow into so beautiful an area and have hit the eye with so

117. 'From a Letter on Petersham Lodge' in John Dixon Hunt and Peter Willis (eds.), *The Genius of Place: The English Landscape Garden 1620–1820* (London, 1975), pp. 148–50.
118. Anthony Ashley Cooper, Earl of Shaftesbury, *Characteristics of Men, Manners, Opinions, Times*, 3 vols (London, 1714).
119. Quoted in Jane Brown, *The Pursuit of Paradise* (London, 1999), p. 61.

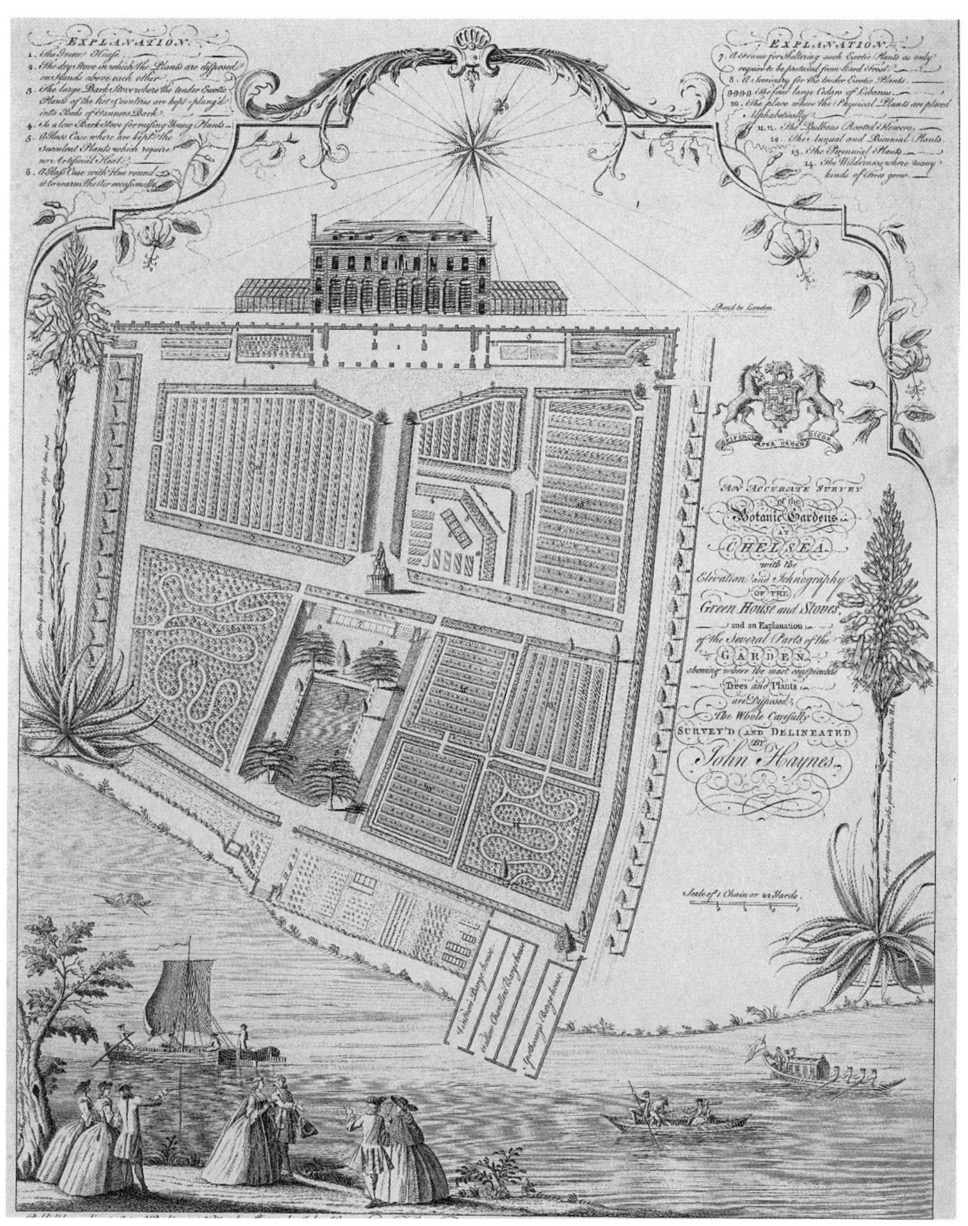

Fig. 12. 'A Plan of the Botanic Gardens at Chelsea by John Haynes', a bird's-eye view of Chelsea Physick Garden, 1751.

Fig. 13. Kensington Palace, *c.* 1724.
© *The Trustees of the British Museum*

> uncommon and agreeable scene as that which it is now wrought into. To give this particular spot of ground the greater effect they have made a very pleasing contrast; for as on one side of the walk you see this hollow basin with its several little plantations lying so conveniently under the Eye of the beholder; on the other side of it there appears a seeming mount made up of trees rising one higher than another in proportion as they approach the centre. A spectator who has heard of this account of it would think this circular mount was not only a real one but that it had been actually scooped out of that hollow space which I have before mentioned.[120]

Despite the large sums of money spent on the garden by King William and Queen Mary, in 1702 Queen Anne had instructed London and Wise to provide 'an English model to the old-made Gardens at Kensington'.[121] For once accompanied by a friend, Molyneux concurs with the queen's dislike of the French Baroque style and, as at Hampton Court and Bushy Park, his appraisal is not resoundingly encouraging. He refers to the transformation

120. *The Spectator*, 6 September 1712, p. 22; Jennifer Ledfors, 'Notes on the Early History of Kensington Palace Gardens', *The London Gardener*, 11 (2005/6), pp. 11–18, and '"The Ingenious Mr Charles Bridgeman" and his work at Kensington Palace', *The London Gardener*, 11 (2005/6), pp. 19–38.
121. Stephen Switzer, *Ichnographhia Rustica*, 1 (London, 1718), p. 83.

of the upper garden, which had been converted from an old gravel pit, and echoes the sentiments of *The Spectator* by commending the mount as being an optical illusion created by varying the height of trees. The gardens, he states, 'are counted a Masterpiece of Art in the new regular manner of greens and gravel gardening', adding, 'I say all this be in its way very agreeable yet in my opinion all this falls so low and short of the sublime unconfinedness of nature [...] I cannot conceive how the world is so entirely fall'n into this way of gardening' (f. **89**). The final words on Kensington Gardens come from Molyneux's unidentified friend, who 'did not like these Epigrams in gardening and was much more pleas'd with the Epick Style or the Pindarick' (f. **90**). This reference to the garden paradise in Greek literature echoes that published by *The Spectator* when, also referring to Kensington, its editor Joseph Addison writes, 'As for my self, you will find, by the Account that I have already given you, that my Compositions in Gardening are altogether after the *Pindarick* Manner, and run into the beautiful Wildness of Nature, without affecting the nicer Elegancies of Art'.[122] This, together with the fact that later on in the letters he refers to being 'sometimes receiv'd by [...] Mr: Addison' (f. **148**), implies that Addison is his companion at Kensington.

Kensington Palace was purchased by King William and Queen Mary in the summer of 1689. Within two years Sir Christopher Wren had added the Queen's Gallery to which Molyneux refers in his account. In 1696, John Evelyn referred to Kensington as '[...] very noble, tho not greate; the Gallerys furnished with all the best pictures of the [royal] houses, of Titian, Raphael, Corregio, Holbein, Julio Romano, Bassan, V:Dyck, Tintoret & others, with a world of Porcelain; a pretty private Library', while later, in June 1733, John Loveday recorded seeing in the 'long room' the Muses by Tintoretto and 'Three beautiful females, perhaps the Graces', which he incorrectly ascribes to Titian.[123] Molyneux correctly attributes the 'Three Graces', now known as *The Toilet of Venus*, to the Italian Baroque master Guido Reni. This painting was given to the nation by King William IV in 1836 and is now in the National Gallery in London. In the 'Prince's Gallery' (f. **90**) (later known as the King's Gallery) Molyneux notes Sir Godfrey Kneller's series of portraits of admirals, nine of which were commissioned by Queen Anne between 1702 and 1714 and later presented to the Greenwich Hospital by King George IV in 1824. The 'four Indian Kings' pictures must have been new when Molyneux saw them as they do not appear in the *c.* 1705–10 inventory of Queen Anne's picture collections.[124] In 1710 four Iroquois leaders — Etow Oh Koam, King of the River Nations; Oh Nee Yeath Ton

122. *The Spectator*, 6 September 1712.
123. de Beer, *John Evelyn*, p. 878; Markham, *Loveday*, pp. 516–17.
124. With thanks to Alex Buck for checking 'The Inventory of Queen Anne's picture at Kensington, Hampton Court, Windsor, St James's and Somerset House' of *c.* 1705–10 (unpublished).

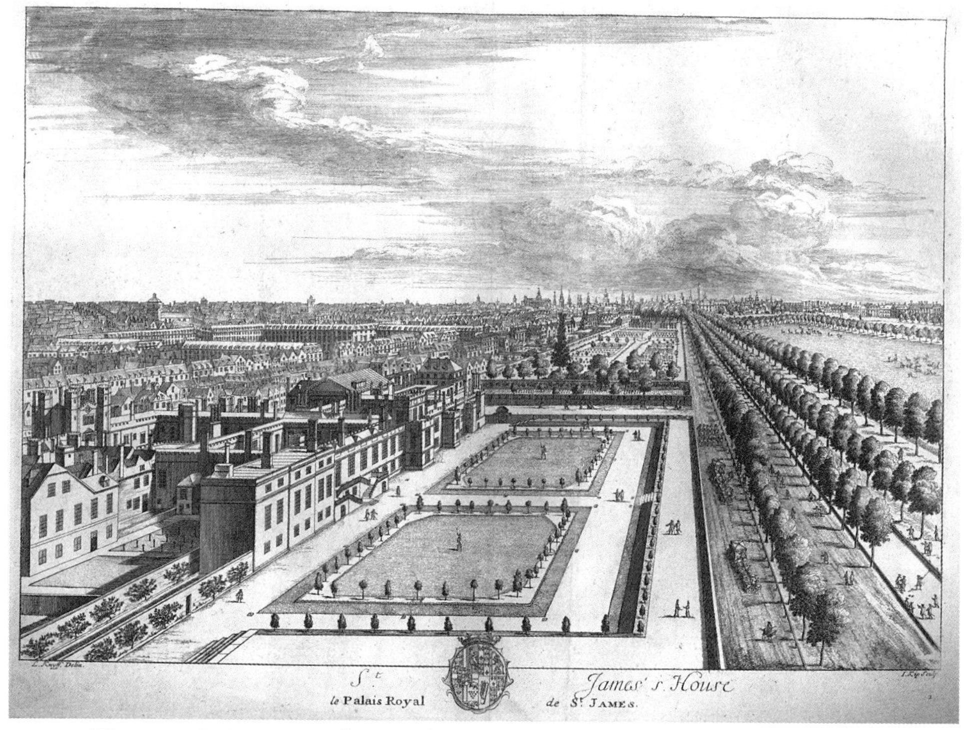

Fig. 14. St James's Palace with the City of London in the background and the Mall on the right, *c.* 1707.

No Rion also known as John of Canajoharie; Sa Ga Yean Qua Rah Ton, King of Maquas; and Theyanoguin, Emperor of Six Nations — visited Queen Anne, an event that aroused great interest in England. Their portraits were painted by Jan Verelst (1648–1734) and were sold in the 1830s from the Royal Collections to the Petre family of Kent who now reside in Canada.

St James's Palace (Fig. 14) (Map V ⑪) and Somerset House receive scant attention from Molyneux. Intriguingly, he had just appeared at St James's Palace as one of Queen Anne's guests at her forty-seventh birthday celebrations on 6 February 1713, although in what capacity he attended remains unclear. His brief account (f. **91**) fits well with Jonathan Swift's summary when, in a 'Letter to Stella', he wrote, 'This is the Queen's birthday, and I never saw it celebrated with so much luxury and fine clothes'.[125] At Somerset House (Map II ⑫) Molyneux is pleased with the river frontage garden (illustrated by Johannes Kip in *c.* 1690) that had 'agreeable' (f. **92**) vistas across the River Thames to Boydell Cuper's garden

125. Scott, *Swift*, p. 425.

and beyond. By 1713 Somerset House had ceased to be a royal residence and was converted into grace-and-favour apartments long before William Chambers rebuilt the palace during the reign of King George III. Molyneux's indifference to the architecture was not reflected by James Ralph when in 1734 he described the buildings as 'the first dawning of taste in England', adding, 'in all probability the architect was an Englishman, and this his first attempt to refine on his predecessors'.[126]

The 'Englishman' alluded to by James Ralph was Inigo Jones, who from 1609 undertook the reconstruction of Somerset House, then known as Denmark House after Anne of Denmark, consort to King James I. Mentioning Inigo Jones (1573–1652) by name (f. **92**), in connection with the Banqueting House, suggests Molyneux's approval of the merits of English Palladianism over the French- and Dutch-influenced Baroque movement which was so prevalent under Wren and John Vanburgh. To declare that Molyneux was a neo-Palladian is certainly a simplification of the evidence; however, it is not unreasonable to suggest that he followed with interest the developments in architectural style just prior to his arrival in London. After all, the beacons which heralded the Palladian revival in England were realised within two years of Molyneux's writing: that is, the first published volume of Colen Campbell's *Vitruvius Britannicus* and the first English complete translation of Andrea Palladio's *Quattro libri dell'architettura*.[127]

Yet, if this was his favoured style, it is remarkable that when at Somerset House Molyneux fails to praise John Webb's palazzo-style Long Gallery, built in 1662 (simply calling it 'old') (f. **91**), or that no mention is given to either the Queen's House at Greenwich (also by Jones) or to Peckwater Quadrangle at Christ Church, Oxford, the earliest Palladian square in Britain. It is possible that there is an element of truth in his confession that 'you see I am not Architect enough to attack St Pauls in form but however if I have not the Art I have not the Envy of a Critick and you are sure to have Nature's judgement unbyass'd in what I say' (f. **35**). His antipathy to the Baroque in both architectural and garden design may well have been influenced by political or religious bias — indeed, the very fact that the Baroque developed within the European Catholic culture may have proved anathema to Molyneux's Whiggish and Protestant sensibilities. Consequently, in his letters only one 'Pallace' amounts to his 'Ideas of Grandeur & Sublime' — the Banqueting House (Fig. 15) (Map VI ⑬) (f. **91**). By calling it a palace Molyneux appears to elevate its status towards being a key element in the rebuilding of Whitehall after the fires of 1691 and

126. Ralph, *Critical Review*, p. 38.

127. Colen Campbell, *Vitruvius Britannicus, or the British Architect* (London, 1715–25); G. Leoni (ed.), *The Architecture of Andrea Palladio* (London, 1715–20). It is of note that Molyneux subscribed to William Kent's *The Designs of Inigo Jones, consisting of plans and elevations for publick and private buildings* (London, 1727).

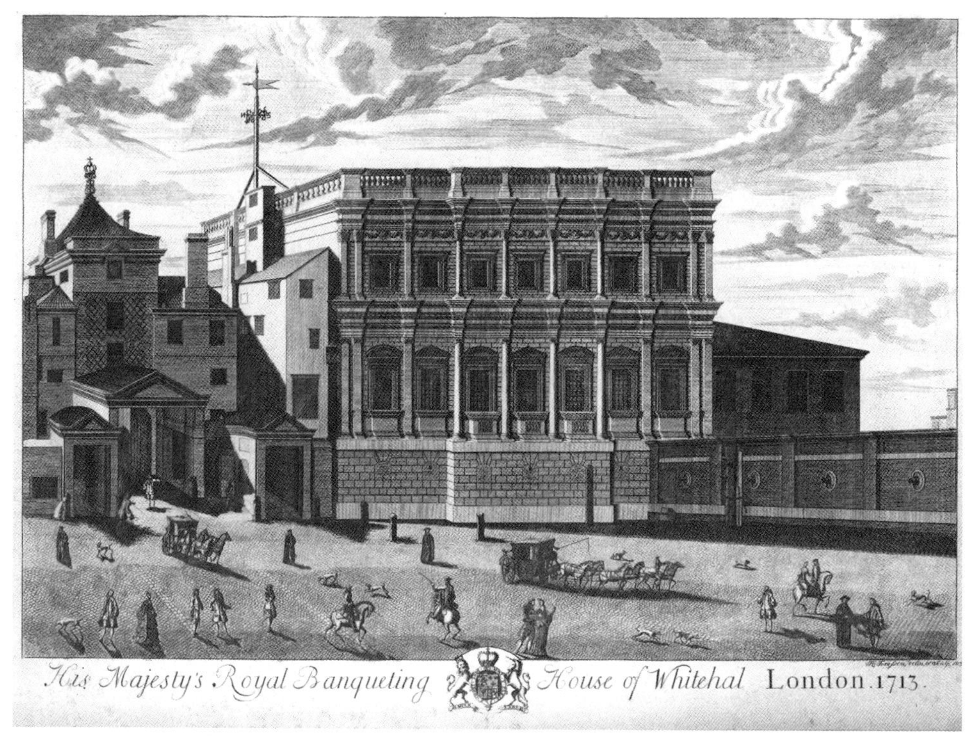

Fig. 15. The Banqueting House in Whitehall, 1713.
© *The Trustees of the British Museum*

1698.[128] His reader, Thomas Molyneux, would have been well aware of the 'dreadful fire' (f. **91**) of 1698, of which a contemporary wrote

> About 4 this afternoon a fire broke out at Whitehall in Colonel Stanley's lodgings near the waterside, which got such a head, that notwithstanding all care that was taken with the playing of Engines, and blowing up part of buildings, the king's apartments, with the lodging adjoining, and part of the new Long Gallery, and the other way the old Presence Chamber, and as far as the chapel, are burned down.[129]

Further avoidance of architectural analysis may be seen in Molyneux's accounts of Buckingham House (Fig. 16) (Map V ⑭), Marlborough House and Berkeley House. The architect Captain William Winde started a new red-brick house for the Duke of Buckingham by 1705, but, despite being hailed by Hatton in 1708 as 'a graceful palace, very commodiously situated [...] not to be contempted by the greatest Monarch', Molyneux modestly

128. John Harris, *The Palladians* (London, 1981), p. 55.
129. TNA SP32/9, f. 6.

Fig. 16. Buckingham House in St James's Park, 1710.
© *The Trustees of the British Museum*

describes it as 'handsome and well worth seeing' (f. **92**).[130] He approves of its situation, in particular of using The Mall as an avenue, a feature also endorsed by John Macky in 1722 and by James Ralph in 1734.[131] The only features mentioned are the wall paintings, presumably those on the staircase painted by Louis Laguerre depicting the story of Dido and Aeneas, which were destroyed by John Nash in the 1820s.[132]

Marlborough House (Fig. 17) (Map V ⑮), although modelled on Buckingham House, is according to Molyneux 'worth seeing not on its own

130. John Harris, 'From Buckingham House to Palace: The Box Within the Box Within the Box', in Robert Simon (ed.), *Buckingham Palace, The Complete Guide* (London, 1993), pp. 28–36.

131. John Macky, *A Journey through England in familiar Letters from a Gentleman to a Friend* (London, 1722), p. 194; Ralph, *Critical Review*, p.180. Of the house, Macky added, 'The staircase is large and nobly painted [...] the apartments indeed are very noble, the furniture rich, and many very good pictures'.

132. The frescoes were copied by James Stephanoff and published in David Watkin, *The Royal Interiors of Regency England* (London, 1984), pp. 80–1.

Fig. 17. Marlborough House, view from The Mall, *c.* 1720.
Private collection

account but for the pictures' (f. **93**). The Duke of Marlborough's London home in Pall Mall, St James's, was rebuilt from 1708 by the duchess and her architects, the elderly Sir Christopher Wren and his assistant Christopher Wren junior.[133] Although Marlborough was a great collector and patron, the contents had, in part at least, been dismantled because the duke and duchess had gone abroad. It is of note that the wall paintings in the Salon depicting scenes of the Battle of Blenheim, again by Louis Laguerre, had been completed by the time of Molyneux's visit but are not mentioned. Another curiosity is his confusion over Van Dyck's painting of King Charles I on horseback, which he misidentifies as King James I. Of this painting the duke wrote to the Duchess of Marlborough on 8 November 1706,

> I am so fond of some pictures I will bring with me, that I could wish you had a place for them till the gallery at Woodstock be finished; for it is certain that there are not in England so fine pictures as some of these,

133. A. Searle, 'Sir Christopher Wren and Marlborough House', *British Library Journal*, 8, part i (1982), pp. 37–45.

> particularly King Charles on horseback done by Vandyke. It was the Elector of Bavaria's, and given to the Emperor, and I hope it is by this time in Holland.[134]

This painting was later transferred to Blenheim Palace, sold in the 1886 by the 8th Duke of Marlborough and is now in the National Gallery. The leather wall-hangings, presented to the duke by Victor Amadeus, Duke of Savoy later King of Sardinia, are said by Molyneux to be by Titian, a fact much doubted by 1766.[135] Later known as *Loves of the Gods*, these hangings along with paintings by Rubens were also moved to Blenheim Palace where they were destroyed by fire in 1861.[136] Today Marlborough House is the headquarters of the Commonwealth Secretariat and Commonwealth Foundation.

Berkeley House (Map I ⑯), home of William Cavendish, the 2nd Duke of Devonshire, was situated on the north side of Portugal Street, near Piccadilly. Molyneux describes the contents of his host's house as 'one of the best collections in England' (f. **93**) but, as with the previous two houses, passes little judgement upon their qualities. If his antiquarian mind is not stimulated by these three conspicuous displays of wealth and connoisseurship, then his next visit to Dr Woodward's museum (Map IV ⑰) must have proved one of the highlights of his time in London.

Dr John Woodward (1665–1728) was a scientist, physician, member of the Royal Society and friend to the Earl of Pembroke, John Locke and Thomas Hearne. Molyneux, like von Uffenbach two years earlier, has a personal tour around the collections held in his rooms at Gresham College (Fig. 18) but, unlike the German, he does not experience the discourtesy of being kept waiting. Von Uffenbach wrote, 'This is the discourteous little ceremony that the affected and pedantic mountebank makes a habit of going through with all strangers who wait on him'.[137] Being an avid collector of antiquities, Dr Woodward shows both men his famed 'shield', an embossed iron buckler which he believed was Roman but is now understood to be of French origin made during the mid-sixteenth century. Woodward acquired from piece from John Conyers (1633–94), an apothecary of Gray's Inn Lane, who, according to the well-reputed book dealer John Bagford (1651–1716), 'made it his chief business to make curious observations and to collect such Antiquities as were daily found in and around London'.[138] Woodward's

134. Quoted in *The Literary Gazette*, no. 1097, 23 December 1837, p. 809.
135. G. Scharf, *Catalogue Raisonné, Blenheim Palace* (London, 1862), pp. 83–92.
136. *Illustrated Times*, 16 February 1861, p. 102.
137. Quarrell and Mare, *von Uffenbach*, p. 176.
138. J. Bagford, 'Mr Bagford's Letter Relating to the Antiquities of London', in Thomas Hearne (ed.), *Johannis Lelandi antiquarii de rebus britannicus collectanea* (London, 1715), pp. lxiii–lxv.

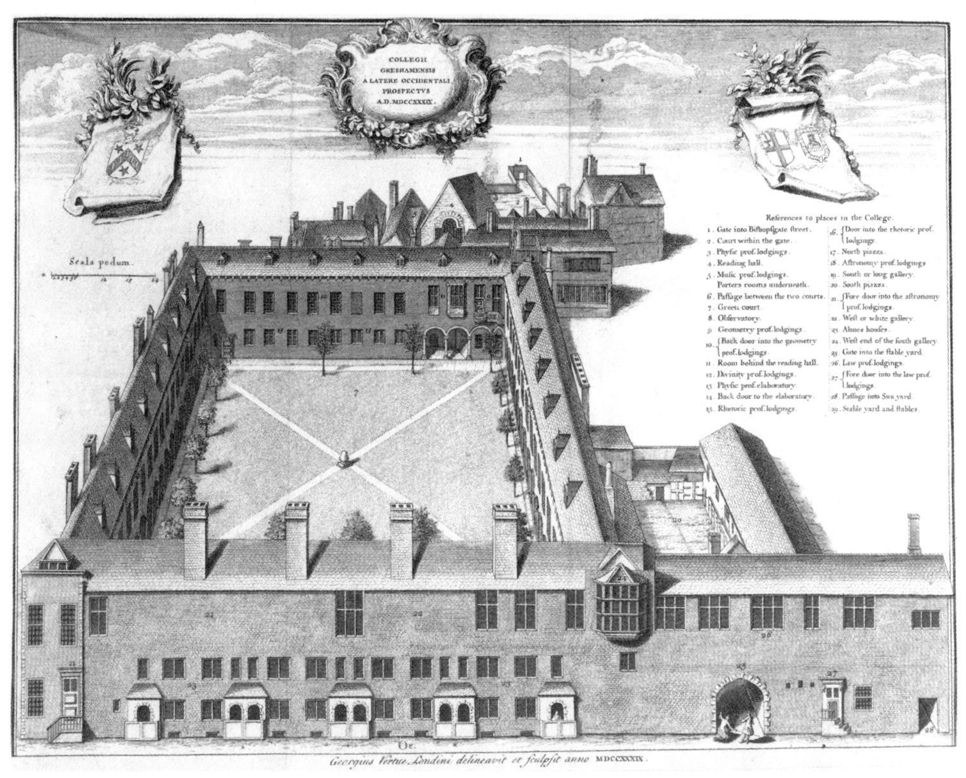

Fig. 18. Gresham College, Dr Woodward's rooms are on the first floor of the rear range, marked no. 12 'Divinity Professor's Lodgings', 1739.
Private collection

1713 treatise on the shield's Roman provenance caused great consternation amongst satirists of the day such as Alexander Pope and Jonathan Swift who discredited Woodward's theories as 'false tastes in learning'.[139]

But, for Molyneux, it is Dr Woodward's geological collections which represent a shrine to learning. Between 1688 and 1724 he amassed some 9,400 scientific specimens which were painstakingly classified and documented.[140] From this collection he formulated intellectual theories, such as *An Essay towards a Natural History of the Earth* (1695), in which he contentiously debated the origins of fossils — then one of the most controversial aspects

139. Levine, *Dr Woodward's Shield: History, Science, and Satire in Augustan England*, p. 2.
140. V. A. Eyles, 'John Woodward FRS, FRCP, MD (1665–1728): A Bio-Bibliographical Account of his Life and Work', *Journal of the Society for the Bibliography of Natural History*, 5, no. 6 (1971) pp. 399–427; John Ward, *The Lives of the Professors of Gresham College* (London, 1740), pp. 283-301.

of natural philosophy.[141] Molyneux's admiration of Dr Woodward's collection is clear, as his account covers more than four folios and provides a comprehensive description of the British and foreign archaeological artefacts, fossils, minerals and rocks on show (ff. **93–7**). Yet, despite Dr Woodward's idiosyncratic charm, Molyneux appears sceptical of his theories; in particular his notion that Noah's flood dissolved the world which eventually rearranged the earth's strata according to the specific gravity of materials.[142] Of this theory Molyneux writes, 'The Doctor however is so well pleas'd with being sure he is in the right that it were a Cruelty if it had not been an impossibility to convince him of the contrary' (f. **96**).

For Molyneux, part of the attraction of this collection would have been the physical arrangement and methodical ordering of Woodward's manuscripts. Amongst prints on ancient buildings by 'La Frere', probably Antoine Lafréry (*fl.* 1544–77), 'Enea Vicho' (1523–67) and 'Mark Antonio' Raimondi (*fl.* 1482–1527), were 'unpublished Manuscripts' compiled by Woodward himself as well as 'several Catalogues of his own Museum and Fossills' (ff. **96–7**). These catalogues were posthumously published and directly correspond with the collections housed in five cabinets.[143] The fact that these cabinets are not mentioned in the letter may well suggest that they were not *in situ* in 1713. After Woodward's death the collection remained intact, a testament to its importance. It was bequeathed to Cambridge University where it formed the founding collection of the Sedgwick Museum of Earth Sciences and where it still remains in its original cabinets.

Text of Letter 3 (complete)

[**83**] Dr Sir Lond:^{n} Febry 14th 1712/3 [144]

You will permit me among the Remarkeables of London to Reckon the Palace of Hampton Court which you know lyes up the River Thames, and so near that you may see it and return the same day very easily; Wee went down, I remember, in a Stage Coach, which Voiture by the way you have from London almost to all parts of the Kingdom and not at dear Rates. At Hampton Court you see a very fine old Palace built by Cardinal

141. Molyneux's interest in fossils is suggested in the Pitt collections catalogue by the presence of Martin Lister's *Historiae sive Synopsis methodicae Conchyliorum*, 4 vols (London, 1685–92).
142. Jill Cook, 'The Nature of the Earth and the Fossil Debate', in Kim Sloane (ed.), *The Enlightenment* (London, 2003), pp. 92–6.
143. D. Price, 'John Woodward and a Surviving British Geological Collection from the Early Eighteenth Century', *Journal of the History of Collections*, 1, no. 1 (1989), pp. 79–95. A catalogue of the collection was published under the title *Catalogue of the Library, Antiquities, etc of the late Dr Woodward* (London, 1728).
144. Refer to 'Manuscript' section for clarification of the dating system.

Woolsey and given by him in his Disgrace to the Crown, the Chappel and Hall of which are handsome enough for so old a Building, and joyning to this you see a new square of Building to the gardens made by King William of about 100 yards front, one of the apartments of which is furnish'd & finish'd, the Stair Case is done by Vario as is the Bedchamber and several other Rooms in the unfinish'd part of the building in as fine, glaring Colours as can be seen in any of his Labours; the greatest Curiosity I observ'd on the Stairs, besides that he represents all the Caesars there alive at once, is the very great Eccho, which is so strong that the lowest voyce is indistinctly [gap] with it. The Guard room which you come into first makes a very beautiful and well dispos'd Armory, the Arms are here hung only against [**84**] the Walls so that in my opinion it made at least a prettyer figure, tho not a more useful one, than that at the Tower; Wee observ'd some of the Boards in this Room at least 60 or 70 feet long in the Floor, from hence you enter the Apartment which fronts one of the Gardens and consists of several well proportioned rooms, a plein pied generally furnish'd with Tapistry, a Few Pictures of Vandyck and some pieces of perspective done by [gap] over the Doors; the Bedchamber has nothing very extraordinarily fine and indeed the Furniture of the whole apartment had so little Effect on me that I had scarce spoke of it but that you pass thro this to a very large Gallery which fronts another Garden and does verily deserve to be seen with great attention on Account of the Cartoons that hang there of Montano, there are nine of them which seem to have been formerly joyn'd all in one representing a Triumph of Julius Caesar as large as the life, the great Number of Figures and the exact Fidelity of the Painter in the Representation of this Triumph and all the Anquititys belonging to it is what makes this work so valueable as it is, for the design of any single figure is by no means correct or extraordinary, I have been told he exactly follows the description of Paulus Emilius his Triumph in Plutarch; if you are curious to See these Designs you will find Prints taken from them in D[r].Clarkes new Edition of Caesars Comentarys. But however these [**85**] are not the famous Cartoons that make so valueable part of the Furniture of Hampton Court they shew you another Gallery where you have the Seven Cartoons of Raphael of which you know the Prints are handed about every where, these you have there so tolerably well represented tho' nothing equal to the Originalls that I need say no more of them, but that they are very well preserved and the Colours inconceivably lively for their Age. While we were there we saw Mr Dorigny who is the famousest graver in the World and is at present employ'd in graving these Cartoons; he shew'd us two plates which are finish'd and are indeed extraordinary, They are twice as large as the Mezzotints and do infinitely more Justice to the Strength of Expression that is remarkeable in Raphaells paintings. I cannot omit telling you that these Cartoons were bought at a prodigious Price by the Advice of Rubens, I think for 70000 [li] and are but a part of a set of Raphaells designs, the rest of which are at Rome and were drawn by him only to work hangings on for the Pope. I think I have now told you every thing I observ'd remarkeable in the Apartmn[ts:]

Here except that I took notice almost every room had got a Weather glass in it, which I suppose was a remain of the Late Poor Kings Furniture. The Situation of Hampton Court is on a dead Flatt and for such a one, I think the Gardens, the great Canall, the Park and the Bowling green are beautifully dispos'd enough, The Thames which runs at the Foot of one of the Gardens and along [**86**] the great Terras going to the Bowling Green makes one of it's greatest beautys, and has given an Agreable Situation to a very handsome banqueting house, which is built close on the Bank of the River; As I have not Botanicks enough to tell you all the Exotick Plants they shew'd us here in two or three different green houses and as I beleive the Comon prints of Hampton Court will give you a better notion of it's Walks, Labyrinths & Scituation than I can if you have not seen this Palace yourself, I shall confine my Self to say no more of it than this, that I did by no means think it adequate in the whole to the notion I had of the Palace of a great Prince; From Hampton Court we went to see a lodgeing Park call'd Bushy Park belonging to the Palace which is now in the hands of my Lord Hallifax as Ranger I believe of that Park, there was little or nothing remarkeable but the Cascade which was not very high, but little and yet very beautifully dispos'd so as to fall between two fine pieces of Grotto work where are places left for Paintings representing two Caves in which the little walks around the Basin of the Cascade end; the Paintings are moveable so as to be taken away in Winter. From hence we went down the River by Boat and saw a Beautiful Country all along a great many Villages and several pretty Seats along the Banks, and in about an hour I think we came to Richmond which is indeed a very pretty village as I have seen from thence we went to see New Park which is [**87**] a Seat just by belonging to my Lord Rochester; if Hampton Court did not fill my Expectation the Gardens here I assure you did pay the Pains of my Journey and gave me perfect Satisfaction. I think I have never yet seen any piece of Gardening that has so much as this the true taste of Beauty, There is a certain sort of Presumption appears in the comon restrain'd formal & Regular Parteres and Gardens that one meets with but here art has nothing Sawsy and seems to endeavour rather to follow than alter nature, and to aim at no beautys but such as she before had seem'd to dictate. The Partere behind the house and the Hornbeam walks beyond it are well enough but a very high hill to the left of the Gardens, part of which is beautifully and wildly dispos'd into Slopes, the rest and upper part cover'd with a fine wood so interspers'd with Vistas & little innumerable private dark walks thro' every part of it lin'd on both sides with low hedges with the unconfin'd Prospects you meet every now and then of the Garden below the Country and the River beyond you is what in my opinion makes the Particular & distinguished preferable beauty of this place – beyond any thing that I have ever yet seen. I should not forget to tell you that in every walk you meet here and there a little opening in the wood with Seats, a Statue, a grass plott, a Basin of water or the like, for you must know that here is a Fountain playing on the side of the hill (the Spring being **[88]** at the Top) which is much higher, the Basin of which lyes higher than the house Roof as well as I remember they shew'd us here a large Cedar of Libanus, in the house I think I observ'd nothing remarkeable at all, From hence persueing our Journey to London you meet Fulham the residence of the Bishops of London whose Garden of Exotick Plants is worth seeing and opposite on the other side Putney a pretty Village. Soon after we landed at the Colledge of Physicians' garden at Chelsea where we saw several Exotick Plants more Particularly four very fine Cedars of Libanus and a Cork tree I took notice on which we saw the Bark perfect good Cork. You see here Chelsea Hospital for maim'd Soldiers, the apartments of which are much like those at Greenwich tho' not near so neat and elegant but for the Corps of the two buildings they are by no means comparable Chelsea having in my Opinion nothing at all extraordinary in regard to the other. Having seen Chelsea we walk'd in London & Wises Gardens and saw there their vast Collection of all Sorts of Greens which are indeed extraordinary and cut into the most Bizare & Topia Figures in the world some representing a Deer, some a man, some a hand and arm, and twenty other Shapes. here crossing the Road you see the Pallace of Kensington the gardens of which it seems **[89]** are counted a Master peice of Art in the new regular manner of greens and gravel gardening; for my part I must confess I have no opinion of this way at all, and tho I own that the gravel pit at Kensington is happily enough dispos'd, the antique Busts and Statues very well plac'd at the end of the little walks, tho the mount made in Appearance on Level ground by trees of Different heights, and the small circular compartment of high greens I observ'd with a Statue in the middle of it of about 40 or 50 yards Diameter with 8 walks centering there and as many seats let into the hedge between them, tho I say all this be in its way very agreable yet in my opinion all this falls so low and short of the sublime unconfinedness of nature, and there is something in infinitely more exalting in the beautiful Scaravagie of noble grown Trees in a wild wood that I cannot conceive how the world is so entirely fall'n into this way of Gardening. The pleasure a great man takes in being able to force nature, and to make and finish a Garden in a Season is what has certainly introduc'd this, whereas I think a well chosen exalted Scituation with natural Wood and Waters & a distant prospect in spight of all these little vain Efforts of mankind does still display greater beauty, and is in its wild variety more inviteing to the noble Seat than this **[90]** in all its finish'd Regularity is pleasing after all; Variety is confess'd to be the Beauty of a Garden, now there can be no variety in that that one sees all the parts of so plainly at once and this seems to be a natural Reason against this kind of gardening; a Friend of mine that walk'd with me there and happen'd to be of my opinion methought say'd a pleasant thing enough on this head, that he for his part did not like these Epigrams in gardening and was much more pleas'd with the Epick Style or the Pindarick. The Lodgeings at Kensington are very well worth seeing on Acc^t of the great number of fine Pictures that abound there being all Originalls of the best hands it were endless to describe each one that pleas'd me, however there is in the Queen's Gallery, the Muses of Tintoret a Madonnas head & a Child a Cartoon or else a Piece in Fresco by Raphael, two or three naked women by Guido I believe that pleas'd me very much and I think more than the rest tho indeed I admir'd everyone I saw extremealy, in the Prince's Gallery you see Portraits of many of the Admiralls of England, and of the four Indian Kings that were here, all by S^r: Godefry Kneller,[145] in the Prince's Closet I saw a Cabinet of Gems and Rings which I believe may be very valueable, but being lock'd up we could not discern them nearer than thro' the glass, I had like to have forgot the hangings in the Queens Bedchamber **[91]** which are indeed extraordinary; when I have mentioned one Palace you will expect to know what I think of another I mean St James's, to tell you the truth I did not think it worth visiting till the Queens Birth night, and then I was there, indeed the Queens presence who play'd at Basset for about half an hour tho' in the Gout and then retire'd, with the great crowds of fine Ladys and fine Cloaths, Jewells and embroidered Velvetts to the number of at least 40 or 50 did indeed divert me in a place where nothing else could have invited me, the Pallace having nothing at all to make it such but spacious rooms. Sommerset house in which by the way the French Ambassador is now settled is just such another, rather worse. I remember I saw there an old long Gallery crowded with old decay'd unfram'd Pictures lying on the Tops of one another of which I could see none extraordinary, tho there might have been many valueable ones there hid behind. To confess a Truth to you, I have now seen Hampton Court and Kensington, St James's and Sommerset house and for Pallaces I never thought I should have seen such in all my Travells; there are none of them in my Opinion sufferable but Hampton Court and that not extraordinary neither. I must own however that there does remain one Relick of a Pallace that indeed amounts to my Ideas of Grandeur & Sublime and this in the Banquetting house all that is left of Whitehall by a dreadful Fire, **[92]** it is now made into a Chappel, the Architecture without is very Beautiful & great, within it is one noble large room, perhaps 100 Feet long, vastly high and perfectly well inlightened, the places in the Wall left for pictures, and the ceiling which yet remains very boldly painted by Sir Peter Paul Rubens

145. The German-born Kneller became Governor of the Academy of Painting in 1711, and in tune with the enlightened times was created a Baronet in 1715. Kneller had painted William Molyneux in *c.* 1696 (National Portrait Gallery no. 5386).

together with the rich hangings with which they told me it was once adorn'd did certainly together make a most noble fine Room fitt for the Purposes to which it was design'd – the Entertainment of Our own and the Reception of Foreign Kings by their Ambassadors; this Banquetting house was built by Inigo Jones (of whom I see no remains anywhere but I like them) and is esteem'd as good a piece of Architecture as any on this side the Alps. The Situation of this and of Somerset house is just on the River, and but a small garden between which to me was very agreeable tho' perhaps not very wholesome; on the other side of the Street you have the Cockpitt, the Horse Guard, the Admiralty & several offices of the Court that are not worth speaking of, but that you pass through them in going to the Park and the Mall, a Diversion much us'd in King Charles his time, but now much neglected, One of the walks of the Mall is contriv'd to serve as an Avenue to Buckingham house which is indeed very handsome and well worth seeing. The Duke has there several very good Pictures especially in the Sallon which is very large and the Walls & roof [**93**] entirely painted in Frescoe, and looks out to the Park and that noble Avenue I spoke of. Near this the Duke of Marlborough's house is worth seeing not on its own account but for the Pictures among which there are several of Rubens, King James the 1st on horseback by Vandyke taken at Munick from the Elector of Bavaria & nine history Pieces on Gilt Leather done by Titian, and these Pictures and all the Furniture of the house which was very fine are taken down since the Duchess and her Lord went off, And here I must not forget to mention Berkeley house now the Duke of Devonshires in which I had the Honour to be shewn by his Grace several most Extraordinary good pieces of the best Masters that deserve mighty well to be seen, and I believe make one of the best collections in England. I think I have room enough left on my Paper to give you here an account of the Museum of Dr: Woodward who you know writt the Theory of the Earth, this is indeed extraordinary and deserves in my opinion as well as any of these Palaces to be visited by a Curious Stranger. The Dr: here shews you a very good Collection – several Bas Relieves, Busts, Urns and Vasai of the Romans almost all found in Brittain of great variety & very perfect in their several kinds, amongs these Antiquitys you see his Shield of which I don't doubt but you have the Print, you see also a Roman Liquid Measure call'd the Congius which is very rare & a small Statue of Diana in Brass found near St Paul's church in London [**94**] When you have seen this Antiquitys he shews you a much greater & more valueable Collection of English Aggats & Pebbles, Mineralls, Fossills & Form'd Stones all of England which are indeed extreamly curious and surprizeing & it is with great Pleasure one may there observe the vast variety and yet very remarkeable agreement there is of all the Oars [Ores] of one and the same Mettal, He shews you several Chrystalls that often accompany these different Mineralls, wherein to each of them you observe Nature has given a different Shape, weight and Colour according to the Oars [Ores] they accompany. Thus those of Iron are red, and of the Colour of the Oar [Ore], those of Copper green & blue, those of Lead yellow & Cubical, and thus he conceives all pretious stones receive from some adjacent mineral their tincture. His reasoning is very just, as to Lead when he assures you that in the clearest Cubical Chrystalls he has found a certain quantity of Lead and that as their Opacity & Weight continues to encrease which they do together, the quantity of Lead is greater & greater 'till at length when entirely Opaque they are perfect Oar [Ore] of Lead. This collection as we have hitherto describ'd it, contains the most Elucidateing Materialls that I have seen to a History of Natures hidden Processes within in the Formation of Mineralls, it wants not as numerous Arguments for the Doctor's particular Opinions as to the Generation of form'd Stones, You see here a very great quantity of them from all parts of England & indeed from [**95**] all parts of the World, some of Flinty & some of other Substances so exactly adapted to the moulds in which they were cast as he pretends at the Deluge, together with those Moulds still perfect Shells in all respects & qualitys whatever, particularly I remember 2 great Lumps of Stone in which there were many hundreds of included perfect Shells & this he assur'd us he had from some part of England where there are vast tracts of Land much of the same Substance that to confess a truth, one that sees his Collection will scarce doubt thus far of his Hypothesis that these form'd stones were once in a State of fluidity & obtain'd their present shapes from these their moulds, But whether these moulds being found in places so remote from the Sea as they are in the Alps & elsewhere be a Sufficient Argument that they were transported there by the Deluge as the Doctor would have it thought especially since he shews you the American Murex and several other Shells from thence with form'd Stones enclos'd and all found in England & the Murex at least 200 feet deep tho' the shells be not now known there & whether any man can believe that the Water of the Deluge or any other cause at that time acting was the Menstruum by which these Stony Substances were rendered fluid, while animal & Vegetable bodys were not at all alter'd as is plain by several Stones with not only the Impression of Fern but even the very plants included 10 which he shew'd us found at least 700 feet depth underground [**96**] whether I say these transportations & Formations of nature to shew that they were perform'd in the time of the Deluge is, as the Doctor in his Hypothesis imagines, this is another point in my opinion & what his Collection as far as I could see did by no means certainly evinces, The Doctor however is so well pleas'd with being sure he is in the right that it were a Cruelty if it had not been an impossibility to convince him of the contrary, I must observe to you that he assur'd us with the same Confidence that your great Moose deers Horns found in Ireland were transported thence by the Deluge from America & never belong'd to any Creature that had life there.

When he had seen his Collection of English Fossils & Form'd Stones he shew'd us several Exotick Specimens of the same and other Raritys among which I remember none very remarkeable but the Plant call'd the Agnus Scythicus. I think he say'd he had it from China; 'tis a very odd one and in the Shape of its fruit, which is cover'd with Cotten, very extraordinarily justifys its name, & these with several prints of antient Buildings by La Frere, Enea Vicho & Mark Antonio and several Drafts of many of his Curiositys, besides a shelf full of his own unpublished Manuscripts do make up this Museum. Perhaps the D$^{r:}$ may take it ill if I don't mention his Works more particularly. For Fear of this therefore I must let you know that he shew'd us among these a new **[97]** method of Fossills, an Apology in Latin for his Theory against Camerarius, a Discourse of the Egyptian Antiquitys, several Catalogues of his own Museum and Fossills, the Progress of knowledge in the most ancient ages of the world near the Deluge, The Art of Mineralls & Essaying and several others that I forget and only wish that for his and the World's sake they may be judicious and methodical commonplaces which I believe is the most we can expect from those performances.

I am Y^{rs} sincerely

LETTER 4

(complete: ff. **98–104**), 18 February 1713

SCA D/M 1/3

Molyneux's next letter (Fig. 22), written from London on 18 February 1713 (only four days after his previous one), deals with two subjects that held a particular power over his imagination — books and astronomy. A drive for knowledge, extending well beyond antiquarianism and the natural world, and the promise of self-improvement through observation and ordering, feature prominently throughout Molyneux's correspondence and are fuelled by his visits to well-established libraries. His passion for scientific research, largely in collaboration with James Bradley, brought to fruition the design and manufacture of practical optical instruments and yielded significant astronomical discoveries. This complete letter recounts Molyneux's response to two places which embody his fascination with both fields of interest: the Harleian Library (Map I ⑱) (which he refers to as the Lord Treasurer's Library) and the Royal Observatory at Greenwich.

During his time in England Molyneux visits the Bodleian and Trinity College Libraries in Oxford, the Cotton collection and Lord Halifax's Library at Westminster, yet it is his account of the Harleian collection that is the most detailed. On 17 February 1713 Molyneux, accompanied by Signor Bianchini, pays a visit to York Buildings, Duke Street, near the Strand, where they are received by the part-time Harleian 'Library-keeper' and co-founder of the Society of Antiquaries, Humfrey Wanley. The collection's founder Robert Harley (1661–1724), the 1st Earl of Oxford, and his only son Edward (1689–1741) are not present on this day. Edward Harley assumed partial care of the collection in 1711 when his father was appointed Lord Treasurer, and eventually took full control in 1715 after Robert Harley's impeachment and imprisonment in the Tower of London. There is no evidence to suggest that Molyneux ever met with either Harley, although Robert is mentioned twice more in Molyneux's letters: once regarding his lodgings in Windsor (f. **105**) and again with reference to the notable gem collection amassed in the seventeenth century by Thomas Howard, the Earl of Arundel (f. **143**).

Molyneux's appreciation of the library is written at a time when Harley had been collecting for seven years. The antiquary, palaeographer and non-juror (one who refused to swear allegiance to King William and Queen Mary), Humfrey Wanley had commenced his *Catalogus brevior* in

1708,[146] took full responsibility for the collection from 1715 and actively collected and catalogued until his death in July 1726. During his visit Molyneux notes 2,000 ancient manuscripts, 13,000 deeds and 1,000 rolls, a figure not dissimilar to Wanley's own total, recorded in 1715, of 3,000 printed and manuscript books, 13,000 charters and 1,000 rolls. By the time of Robert Harley's death there were 6,000 volumes of manuscripts and 14,500 charters and rolls, and in 1741, at the time of the 2nd Earl's death, the collection totalled 50,000 volumes, 400,000 pamphlets and 41,000 prints.[147] Molyneux is shown some of the greatest treasures from the collection and it would appear that he copies down sections of Wanley's catalogue for the benefit of his reader. Indeed, so clear are these descriptions that nearly all of the books listed are easily recognisable in the Harleian collection today.[148] The library was purchased for the nation in 1753 and installed in the British Museum until the creation of the British Library.

Molyneux was equally at ease with all levels of society, but it must have been with some trepidation that he visits, on the morning of the 18 February 1713, his father's former friend, the quarrelsome, uncompromising and strong-minded John Flamsteed (1646–1719). Appointed in 1675 by Charles II as the first Astronomer Royal, the dour and perpetually poorly Flamsteed lived in a grace-and-favour house at the Royal Observatory at Greenwich. He and his old friend William Molyneux had quarrelled over the latter's book *Dioptrica Nova*; indeed, so jealous was Flamsteed that Molyneux described him as 'a man of so much ill-nature and irreligion'.[149] Although Flamsteed and the young Molyneux had corresponded, it is likely that this is their first meeting.[150] Their mutual passion for astronomy seems to overcome any tension and Flamsteed allows him the use of his mural quadrant, a telescope fixed to a sturdy wall within the observatory and aligned with the meridian (an imaginary line through the poles). The observer pivots the telescope along the graduated scale until the target is in view, when its position can be measured using the highly accurate divisions on the quadrant (Fig. 19).

146. BL, Add. MSS 45701–7.
147. C. E. Wright, 'Humfrey Wanley, Saxonist and Library-Keeper', *Proceedings of the British Academy*, XLVI (1960), pp. 99–129; Wanley's diaries, covering the period between 1715 and 1726, are in the British Museum, Lansdowne Manuscript, 771, 772 and are published in two volumes: C. E. Wright and R. C. Wright (eds.), *The Diary of Humfrey Wanley 1715–1726* (London, 1966). See also Edward Miller, *That Noble Cabinet: A History of the British Museum* (London, 1973), pp. 45–6, and C. E. Wright, 'Portrait of a Bibliophile VIII: Edward Harley, 2nd Earl Oxford, 1689–1741', *The Book Collector*, 11 (1962), pp. 158–74.
148. Holden, '"One of the most remarkable things in London"'. In this article, books listed by Molyneux are cross-referenced with the Harley catalogue or manuscript number, see footnote 2.
149. K. Theodore Hoppen, *The Common Scientist in the Seventeenth Century; a Study of the Dublin Philosophical Society, 1683–1708* (Virginia, 1970).
150. SCA Molyneux-Flamsteed correspondence.

Fig. 19. Greenwich Observatory showing the Camera Stellata, 1679.
© *The Trustees of the British Museum*

Molyneux alludes to the controversy relating to Flamsteed's *Observations*, written between 1676 and 1705. In 1712 a private edition of Flamsteed's research, entitled *Historiae coelestis*, was published without the author's permission under the auspices of the Royal Society's president, Sir Isaac Newton (Fig. 20), and edited by its secretary, Edmund Halley. So enraged was Flamsteed by Halley's abridgement and unauthorised additions to the text that by 1714 he obtained 300 copies of the print run of 400, removed all 97 pages of which he approved and destroyed the rest. His approved, corrected and expanded edition was published posthumously as *Historiae coelestis Britannica* (London, 1725). The portrait of the Danish astronomer and reformer of observational astronomy, Tycho Brahe (1546–1601), seen here by Molyneux, was displaced by another given by James Hodgson in 1752, now in the Museum of the History of Science in Oxford.

On leaving, he mentions the Royal Hospital for Seamen (Fig. 21) by Sir Christopher Wren, but curiously does not visit it, particularly in the light of the recent completion by James Thornhill of the Great Hall ceiling mural, painted between 1707 and 1712. He also refers to Queen Elizabeth's birthplace, the Palace of Placentia, which was largely demolished during the

Fig. 20. Sir Isaac Newton (1648–1727) after Sir Godfrey Kneller, *c.* 1712.
© *NTPL/Derrick E. Witty*

seventeenth century but of which remains must still have been standing. Charlton House, also in Greenwich, was the early seventeenth-century home of Sir William Langhorn (d. 1715), who it would appear was one of several noblemen who attempted to grow grapes on their estates during the eighteenth century. Today Charlton House is a public amenity run by the London Borough of Greenwich.

Text of Letter 4 (complete)

[**98**] Dear Sir London Feb: 18: 1712/3[151]

Of the Few Curiositys which I have left undescrib'd in London, one that I had the fortune to see yesterday & the other today do very well deserve I should give you a particular account of each of them. What I saw yesterday was the present Lord Treasurer's Library, and this indeed is one of the most remarkeable things in London, It consists almost entirely of Manuscripts of which I believe the greatest Number relate to the State affairs of these kingdoms however not without a great many Curiositys of other Sorts. The Principal of these I took down on my book and shall name them to you, Wee saw an Irish Testament about 700 year old, A Latin one

151. Refer to 'Manuscript' section for clarification of the dating system.

Fig. 21. A view of the Queen's House at Greenwich by Hendrick Danckerts (*c.* 1625–79). Molyneux does not point out the Queen's House in the foreground but briefly mentions the Royal Hospital for Seamen, formerly the King's House, beyond. The remains of the Tudor palace are standing alongside.

© NTPL/Ian Blantern

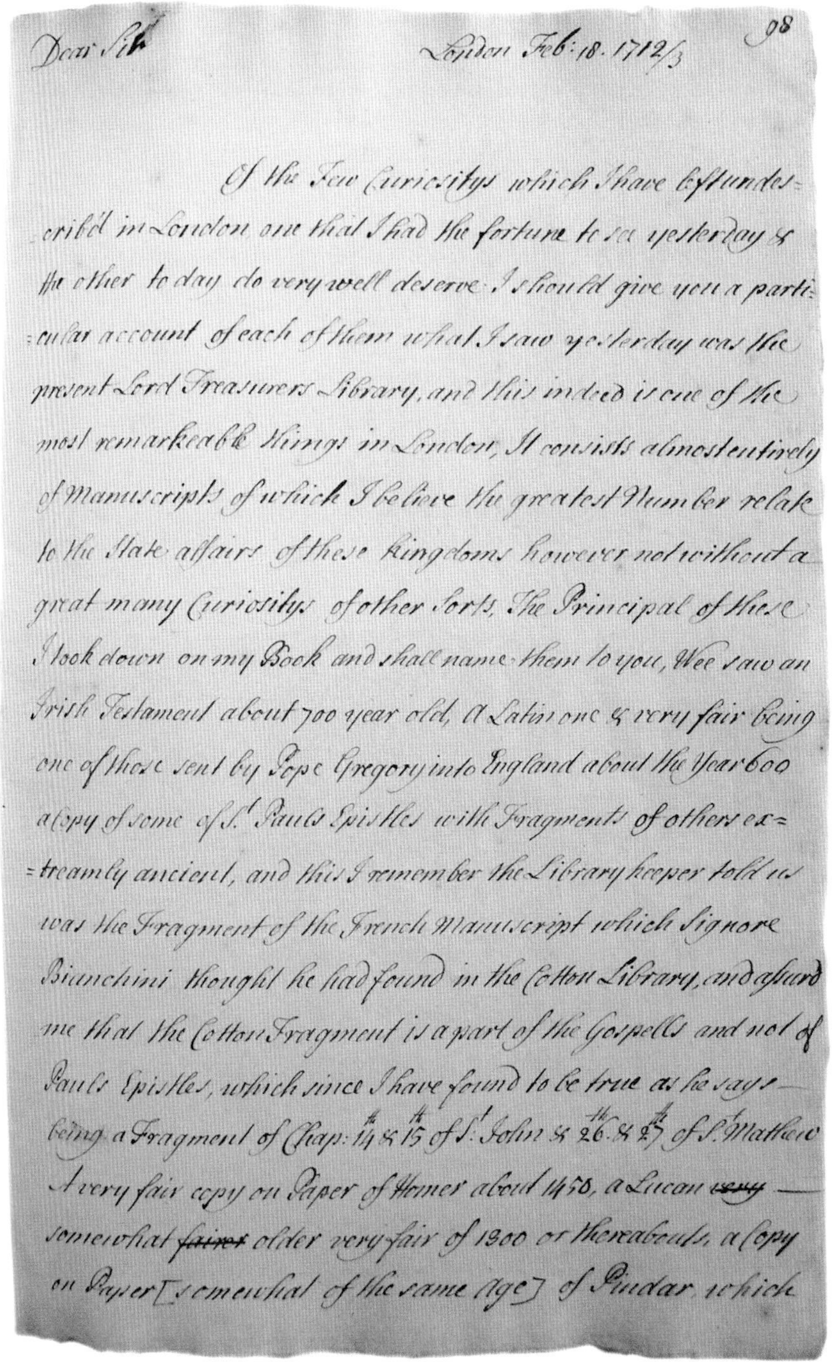

Dear Sir London Feb:18. 1712/13 98

Of the Few Curiositys which I have left undes-
cribed in London one that I had the fortune to see yesterday &
the other to day do very well deserve I should give you a parti=
=cular account of each of them what I saw yesterday was the
present Lord Treasurers Library, and this indeed is one of the
most remarkeable things in London, It consists almost entirely
of Manuscripts of which I believe the greatest Number relate
to the State affairs of these kingdoms however not without a
great many Curiositys of other Sorts, The Principal of these
I took down on my Book and shall name them to you, Wee saw an
Irish Testament about 700 year old, A Latin one & very fair being
one of those sent by Pope Gregory into England about the Year 600
a Copy of some of St Pauls Epistles with Fragments of others ex=
=treamly ancient, and this I remember the Library keeper told us
was the Fragment of the French Manuscript which Signore
Bianchini thought he had found in the Cotton Library, and assurd
me that the Cotton Fragment is a part of the Gospells and not of
Pauls Epistles, which since I have found to be true as he says
being a Fragment of Chap: 14th & 15th of St John & 26th & 27th of St Matthew
A very fair copy on Paper of Homer about 1450, a Lucan ~~very~~
somewhat ~~fairer~~ older very fair of 1300 or thereabouts, a Copy
on Paper [somewhat of the same Age] of Pindar which

Fig. 22. Molyneux manuscript, f. 98, letter dated 'London Feb:18. 1712/13'.

& very fair being one of those sent by Pope Gregory into England about the Year 600, a copy of some of St Pauls Epistles with Fragments of others extreamly ancient, and this I remember the Library keeper told us was the Fragment of the French Manuscript which Signore Bianchini thought he had found in the Cotton Library, and assur'd me that the Cotton Fragment is a part of the Gospells and not of Pauls Epistles, which since I have found to be true as he says being a Fragment of Chap:14th & 15th of St John & 26th & 27th of St Matthew A very fair copy on Paper of Homer about 1450, a Lucan somewhat older very fair of 1300 or thereabouts, a Copy on Paper (somewhat of the same Age) of Pindar, which [**99**] they pretend to be the fairest in the world, together with Orpheus Lycophroim, Homer's Hymns and some other Greek Poets never printed. An Observation the Library keeper made on this Copy is what makes it extreamely valueable and that is, this, that he finds at the beginning inscrib'd the words *y T u* or Bona Fortuna, which it seems was an old Pagan Motto and often plac'd in the beginning of their most ancient Manuscripts, from whence he concludes this must have been a copy of a Manuscript of the greatest Antiquity. The most remarkeable Curiositys he shew'd us relateing to these Kingdoms were, the Charter of Battle Abby an Original, founded by William the Conqueror on the Famous Field of Battle where he gain'd England, A Grant of Coin=gulphs[152] one of the Kings of Mercia dated Anno 814, these two ancient Parchments I observ'd were subscrib'd both of them by the names of all the Nobles and Magnates of the Realm as well as by the King, but all plainly in the Clerk's hand, so that this must be either a Copy or else they none of them could write at all, they are sustain'd to be both Originals but I thought the writing not so old; A more perfect and beautiful Original my Lord certainly has of H:7th Charters of the foundation of his Chappel in Westminster Abby the Privilges of which are as extensive and full as those of some Colleges in Oxford; there are four of these Charters & a fifth uniting them all together with as many Seals under the Great Seal of England, they are mostly neatly bound as they [**100**] were at first in one Volume in red velvet with the Royal Arms neatly enambled on the outside and are as fresh as the first day, I must not forget here a Seal he shew'd us in green Wax mill'd round with Letters near 300 years old belonging to a Deed of the Bishoprick of Norwich, which seal the Bishop still has & uses. And of these I remember Sir Andrew Fountain shew'd me another much of the same Age. For Miscellaneous & more modern Curiositys we saw a French poem in Manuscript relating to the History of Rich: 1st together with History Paintings therein better than could be expected in a Work 400 years old. Another exceeding richly gilt containing in little Oval Gothick Pictures the History of the new Testament somewhat older than the other, a Book of Representations somewhat Mysticall of several of the Popes done about 1431 & extreamely well for that time, a

152. Cenwulf, King of Mercia.

Paper Manuscript in High Dutch with vast number of Figures of Encampments, Marches, Engines &:c relateing to the Art military since the Invention of Guns, but it is a vain attempt I must confess to pretend by a Few we light on here and there to give any true notion of all the Curiositys that my Lord Oxford has here got together. I believe I shall do him better Justice by endeavouring to name no more particulars and by letting you know that certainly as he is a curious & learned Man himself he has employ'd one in this Collection that is equally so too one [**101**] Mr Wanley who is now his Library keeper, and this so successfully that Mr Wanley assur'd me in about 7 years which is the utmost time of their Collection they have got together above 1000 Rolls, 13000 Deeds, and 2000 other Antient Manuscripts, and now have in their Presses within the one single room that makes the Library the entire furniture almost Monasteryes. The way that I have spent this day is I think no less diverting and deserves as well that I should let you know it as yesterday: I went then this Morning to wait on old Dr Flamstead at the Observatory of Greenwich, you know the Friendship there was for sometime maintain'd between him and my Father. On this Account he receiv'd me very kindly, and did me the Favour to permit me to take the Suns Meridian Altitude there by his Mural Quadrant, which I found this 18th day of Febry 1712/3 exactly 59.28'.02" that is of the inferiour Limb, and that the whole disk of the Sun pass'd the threads in 2':17" of time. This instrument with another moveable Quadrant in the Camera Stellata is all he has now in Order or makes use of. He shew'd me several Telescopes which sometimes he uses, of which the Longest was not above 27 Feet and invited me to pass any night there I pleas'd, He shew'd me a Catalogue of the fix'd Stars which he has printed at his own Expence from his Observations much more corrected and compleat than that one printed in the new [**102**] Volume of his Works, which Catalogue it seems he disapproves of and disowns, this he designs not to publish yet awhile, but I believe when he does it will be found at least as Exact and as well accommodated to the ancient Greek names of Ptolemy & to Bayerus his Tables as that in his Works if not better by the way he shew'd 12 or 14 Curious Draughts of as many of the noted Constellations with the Stars in their Scituations plac'd on them in which I really think he was very judiciously corrected, several improprietys in those Tables and on a much larger Scale on this head. Happening to Speak of Sinical Curves he shew'd me a Demonstration he has made about the Sum of the Sines which he says is equal to the Square of the Radius, He shew'd me a note out of Proclus (a very rare book) proveing that there were before Euclid several more eminent Mathematicians than are generally known, one particularly who writt c/c Limitibus Problematum; 'tis a remarkeable Passage & he promis'd me to give it me transcrib'd together with his tables for 1713 of Jupiters Sattelits, and the moons Southing He shew'd me a Picture in Colours a half length of Tycho Brahe and told me one Secret which I must remember in glass grinding & that is, that a glass is seldom good that is polish'd with a Rag and Putty in the hand without the Tool, but that the best and Surest way which he never knew fail was to make the Workman continue to Polish it in the very Tool, and without [**103**] Putty or any other thing but the very Stuff us'd in Grinding which by little must be taken away, and the tool now & then gently breath'd on to make the glass slip about in Surface to which, if the Mould and glass be true, it will stick so hard as to require a good Strength to move it about: Haveing promis'd to waite on him again some night I took my leave & retir'd but cannot but observe to you the Beautiful & Extensive Prospect you have from the Observatory, The fine green park & Deer, the Walks and Trees therein, the Seamen's Hospital and the Village just below you, where by the way you see a poor little house in which Queen Elizabeth was born, together with the River and its Ships Sailing by in view for many miles, the Citty of London to the left and beyond the River an unbounded Plain close set with Villages, Improvements, and these all view'd from a Fine and Pleasant Hill make as agreeable a prospect to the Eye of Sense as 'tis to the Eye of Reason, to reflect that the same exalted Scituation has given as unconfin'd a view of the Heavens, and I could not but recollect with some pleasure the Antient Fable of Atlas, when I found myself standing on a Hill, that may so justly be said to support the Heavens & its motions. From hence we went to see a Seat of Sr Wm Langhorn's in Charlton near Greenwich which is indeed a fine old building & has very large Gardens but I had not mentioned it however were it not to tell you that I there [**104**] drank wine made of English Grapes, It was of a whitish brown Colour, tasted sweet, and I believe had been made with a good proportion of Raisins in it; it was Six years old as they say'd and keeps much longer, The Servant told us they make generally Six or 7 hogsheads in a Season and of different colours out of the Grapes of these Gardens. You will excuse the shortness of this Letter for I am in some haste.

I am Y^rs^

LETTER 5

(two fragments: ff. **105–12, 118–46**), 28 February 1713

SCA D/M 1/3

By mid-afternoon on 19 February 1713, Samuel Molyneux arrives at the small Berkshire town of Windsor. After an overnight stop he takes the road to Oxford, passing through Nettlebed, crossing the River Thames at Henley and travelling through the villages of Ewelme, Benson, Dorchester and Sandford. The account of the nine-hour journey (significantly quicker than the 'flying coach' that took thirteen hours between Oxford and London in 1699), though brief, contains plenty of significant detail. For example, the poor roads are remarked upon, as is the potential threat from highwaymen. Whilst travelling, Molyneux observes the gentlemen's seats of 'Mr Pall', Lady Place (the home of John Lovelace, the 5th Baron Lovelace of Hurley) and the manor house and seat of the barrister Sir James Etheridge at Great Marlow. He also notes some topographical and geological features in the landscape, deferring to the authority of *The History of Oxfordshire* (1677) by Dr Robert Plot (1640–96), the first Keeper of the Ashmolean Museum and Oxford's Professor of Chemistry between 1683 and 1691. Plot's work is also referred to later in the letter.

While staying in Windsor, Molyneux visits Windsor Castle, a royal residence much altered by King Charles II and by now neglected, in which state it remained until King George III chose it as his favourite seat. Like Daniel Defoe twelve years later, Molyneux gives some poignancy to the castle by evoking an idealistic vision based on its Gothic architecture and medieval traditions. This romantic approach, centred around the building's continuity with the past, was popularised by eighteenth-century antiquarians and poets such as Alexander Pope (1688–1744), whose poem *Windsor Forest*, written to celebrate the Treaty of Utrecht which was eventually ratified on 11 April 1713, was published on 7 March (three weeks after Molyneux mentions the poem in his letter). Molyneux's admiration extends beyond the sumptuousness of the interiors to the beauty of the gardens, park and vista (Fig. 23); indeed, he proclaims Windsor to be 'the only Pallace that comes near the Ideas of Pomp and Superiority I have long given in my mind to the Dwelling of a Prince' (f. **110**).

If medievalism holds power over Molyneux's imagination, the Baroque style clearly does not. In his account there is no mention of the seventeenth-century building phases that included Nicholas Stone's 1636 park gateway,

Fig. 23. Windsor Great Park by Thomas Sandby, 1721. Anglesey Abbey, The Fairhaven Collection (The National Trust). © *NTPL/John Hammond*

Hugh May's remodelling of the Upper Ward, St George's Chapel and the King's Chapel carried out between 1675 and 1684 (mentioned in John Evelyn's diary) and Sir Christopher Wren's new guard house built in 1685. Moreover, as at Hampton Court, Molyneux appears somewhat cavalier in his dismissal of the thirteen Baroque-style ceiling paintings (of which only three remain) by Antonio Verrio (1639–1707) that in 1685 were considered as 'stupendious' by John Evelyn, who added that the paintings

> [...] both for the Art & Invention deserves the Inscription, in honor to the Painter *Signior Verrio*: The History is Edw: the (3rd's) receiving the black-prince, coming towards him in a Roman Triumph &c. The whole roofe, the Hist: of St George, The Throne, the Carvings &c are incomparable, & I think equal to any & in many Circumstances exceeding any I have seene Abroad:[153]

That Verrio's unrestrained reading of the classical idiom is not to Molyneux's taste is confirmed in St George's Hall when he describes the murals depicting the Black Prince's triumphal reception by King Edward III, commenting that 'The Subject gave me more Pleasure than the Execution' (f. **107**). More appealing, however, are the wall paintings of *c.* 1700 by Godfrey Kneller, showing King William III '[...] painted on a Throne rais'd on Steps which indeed are perfectly well done when view'd from the true point of Perspective in a Gallery at the other end of the Hall' (f. **107**). Both sets of murals were largely destroyed by the architect Jeffry Wyatville when he joined the old Hall with the Chapel in 1829.

Molyneux's account of the collections at Windsor remains useful as a prelude to the first guidebook that appeared two decades later.[154] Of the highlights, he lists the 'several most incomparable heads by Titian' as being 'the best pictures I ever saw in my life' (f. **106**). Yet the modern reader should exercise some caution when reading Molyneux's identifications. For example, the 'Portrait of Johannes Duns Scotus the famous Philosopher' (f. **106**) is now identified as *A Philosopher Writing* by the Master of the Annunciation of the Shepherds, while the 'History piece [...] of Sr Thomas Gresham buying the Famous Jewell from the Jew [by ...] Quintin [Massys]' (f. **106**), listed in Queen Anne's picture inventory of *c.* 1710–12 as *Two Jews at half-length* is now referred to as *The Misers*.[155] Only brief mention is

153. de Beer, *John Evelyn*, pp. 738–9.
154. Bickham, *Hampton-Court*.
155. Michael Levy, *Later Italian Pictures in the Collection of Her Majesty The Queen* (Cambridge, 1991); *The Misers* attributed to a follower of Marinus van Reymerswaele (active 1535–45) was in the 'Old Gallery' at Windsor where it remained until 1955 when it was tranferred to Hampton Court. Unpublished manuscript of an Inventory of Queen Anne's Pictures at Kensington, Hampton Court, Windsor, St James's and Somerset House *c.* 1710–12 (detail supplied by Alex Buck); Lorne Campbell, *The Early Flemish Pictures in the Collection of Her Majesty The Queen* (Cambridge, 1985), Cat. no. 72, pp. 114–18.

made of St George's Chapel, the spiritual home of the Order of Chivalry founded by King Edward III in 1348. The guard chambers adjoining the hall were later described by Daniel Defoe as having '[...] walls furnished with arms, and the King's Beef-eaters, as they call the Yeoman of the guard, keep their station, or, as it may be called, their main guard'.[156] Here Molyneux reports on preparations for the forthcoming investiture into the noble Order of the Garter of various members of the aristocracy (f. **107**).[157]

Of the gardens, Molyneux's account compares well with contemporary illustrated depictions such as *The North Prospect of Windsor Castle* (*c.* 1704–5) by Leonard Knyff (1650–1722) and the illustration by Johannes Kip (1653–1722) published in *Britannia Illustrata* (1709). However, Molyneux states incorrectly that the 'longest front' was to the east, while in fact the formal garden situated between that castle and river, often referred to as the Maastricht garden, was to the north.[158] The military-style garden (described by John Evelyn in 1674) is, at the time of Molyneux's visit, in the process of being reorganised by Henry Wise to include topiary, fruit trees and vines.[159] Unusually, Molyneux makes no reference to this work other than to note that the garden is 'extreamly steep [and] is cut into 3 or 4 noble slopes' (f. **108**).

Frustratingly, as the coach enters 'the Famous Citty of Oxford' (f. **112**) the subsequent five folios are missing. When the narrative rejoins at folio **118**, Molyneux writes, 'I think I have now told you most of the remarkeable things I saw in the Private Colleges', which indicates that the missing folios contained some reference to Oxford colleges, including those which he later seems to be mentioning for a second or subsequent time: Trinity College Chapel, Balliol College Library and University College Hall. Another college he may well have visited is Corpus Christi where the cloisters were rebuilt between 1706 and 1712, and plausibly the location of the 'Square Pillars' (f. **118**) that the Duke of Marlborough appears to have tried to purchase as architectural salvage for Blenheim Palace. Generally uninspired by the buildings in comparison with those of his native Dublin, Molyneux commends Trinity College Chapel, rebuilt between 1691 and 1694, but unfortunately does not name the architect, who remains unidentified to this day.

It would have been expedient for Molyneux to show restraint in Oxford — a city that remained indifferent to the Hanoverian succession and suffered

156. Furbank, Owens and Coulson, *Defoe*, p. 129.
157. Those being invested as Knights of the Garter on this occasion were Henry Somerset, the 2nd Duke of Beaufort; Henry Grey, Duke of Kent; John, 1st Earl Poulett; Robert Harley, 1st Earl of Oxford and Earl Mortimer; Thomas Wentworth, the 1st Earl of Strafford; and Charles Mordaunt, 3rd Earl of Peterborough.
158. For more on the gardens, refer to Jane Roberts, *Royal Landscape* (New Haven and London, 1997).
159. de Beer, *John Evelyn*, p. 540; David Green, *Henry Wise, Gardener to Queen Anne* (Oxford, 1956), p. 80.

from recurrent Jacobite riots. Yet during his visit to the Bodleian Library his usually polite temper is severely tested by the library's second librarian, the non-juror Thomas Hearne, whose waspish diaries record the visit in great detail.

> On February in the 20th in 1712/13 it being Friday, a little before 10 Clock came to the Library, one Mr. Mollineux, an Irish Gentleman, accompanied by Mr. Keil, our Savilian Professor of Astronomy, and three other Gentlemen, one of which was Mr. Med[l]icot of Westminster. Dr. Hudson was then in the Library, to whom this Mr. Mollineux was recommended by Sr. Andrew Fountaine; but the Dr. being otherwise, perhaps, ingaged had very little Discourse with them. Mr. Keil therefore desired me to shew them the Curiosities of the Place, tho' without telling me, or giving me the least hint, that Mr. Mollineux was a Person of but ill Principles, nor indeed did he so much as tell me what his Name was. I always look'd upon Mr. Keil as a very honest Gentleman, and I knew one of the others to be such, and therefore could not imagine that any one of the company should be quite otherwise.[160]

The first issue here is the date which, it appears, Hearne got wrong. If Molyneux leaves Windsor on Friday 20 February, then he cannot be at the Bodleian Library on the same day, as Hearne indicates in a marginal note in his diary for that date, writing that, 'This Day Mr Mollineux came to the Library before 10.clock and staid 'till past 11 with him Mr Keil, Mr Medlicot, Le. Hunt'.[161] Based on this scribbled note and the fact that his account of the events were written sometime after Molyneux's visit, it is conceivable that Hearne is wrong and not Molyneux, whose dates are consistent throughout.[162]

Hence on Saturday 21 February and armed with a letter of introduction from Sir Andrew Fountaine to John Hudson (1662–1719), classical scholar and Bodleian Librarian between 1701 and 1719, Molyneux's visit to the Bodleian Library starts on a congenial note.[163] His companions that day are John Keill (1671–1721), natural philosopher, Fellow of the Royal Society and, curiously, Decipherer to Queen Anne from 1712, Thomas Medlicott, Deputy High Steward of Westminster, and 'Le. Hunt' possibly Richard Le Hunt MP for Enniscorthy, County Wexford, from November 1713. Considering their guide Thomas Hearne to be 'a very learn'd Man & to whom the World is much obliged for several fine Editions', the party enthusiastically tours the Bodleian

160. Rannie, *Hearne*, pp. 108–9.
161. Ibid., p. 86.
162. Thanks to Kate Dinn for pointing this out.
163. For further insight into the period, see Theodor Harmsen, 'Bodleian Imbroglios, Politics and Personalities, 1701–1716: Thomas Hearne, Arthur Charlett and John Hudson', *Neophilologus*, 82 (1998), pp. 149–68.

collections, taking great pleasure in books and manuscripts singled out for their special attention. Of these, Molyneux notes *On Animals* by the Byzantine poet Manuel Philes (*c.* 1275–1345) (Bodleian Library, MS Auct. F.4.15); a poem owned by the reputed Parisian printer Robert Estienne (1503–59, also known as Robert Stephanus or Stephens); the manuscripts of the English antiquary and topographer John Leland (*c.* 1506–52) (Bodleian Library, MSS Top. gen. c.1–4, e.8–15) and the *Acts of the Apostles* in Latin and Greek (Bodleian Library, MS Laud.Gr.35) dating to 600 AD and said to have been used by the Venerable Bede. On seeing a manuscript of the Old Testament with oval pictures he refers back to the visit he made a few days earlier to the Harleian Library in London (ff. **98–101**), reminding his reader of similar documents described in his previous letter.[164] From his diary entries it is clear that Hearne is unimpressed by Molyneux's observations on the 'Curiosities' he has been shown and begins to question his visitor's scholarly acumen.

From the Bodleian Library, the party moves to the Anatomy School where Hearne parades more fully his 'impertinence' (f. **124**) and political subversiveness. First he attempts to provoke Molyneux by ridiculing the staunch Whig and low-church rector of St Peter-le-Poer in St Paul's diocese, Benjamin Hoadly (1676–1761), who had risen to national prominence in the wake of Revd Henry Sacheverell's disgrace (see Letter 6). Like Sacheverell, Hoadly too became the subject of satirical prints, such as the 'Silly cut of Mr Hodly with horns & asses Ears and they tell you his Wife gave the present' (f. **123**), which Hearne mischievously shows them. However, considering that the print had been in circulation for at least two years, it is hard to believe that Molyneux would take offence. From Hoadly, Hearne's acerbic tongue homes in on a subject closer to Molyneux's heart — the Duke of Marlborough. Hearne recalls,

> And yet for all that he did not discover any Passion, nor give the least Hint that he was a Whig himself. Neither did he give any hint of it afterwards 'till I came to mention a Tobacco Stopper tipped with Silver, and given to me by a Reverend Divine, who had informed me that it was

164. BL, Harley 1526, 1527. The documents referred to are described as follows in *A Catalogue of the Harleian Manuscripts in the British Museum*, 1, no. 1 (London, 1808), p. 9: 'Two very noble Biblical Books, upwards of five hundred years old; being Part of a most richly illuminated Manuscript, the first Volume thereof, beginning at *Genesis*, and ending with *Job*, is preserved in the *Bodleian* Library (Arch. A.154). They consist of Texts according to the Vulgar *Latin*, selected from the Books of the *Maccabees* and *New Testament*, with the subject of each Text, represented in an illuminated Picture, included in a pretty large Circle placed opposite thereto. Underneath each Text is likewise set down, in *Latin*, the Meaning of the same, according to the Opinion of the Author, who generally applies such Text to demonstrate the Benefits of a good Life, and the Punishments attending a bad one. These Explications are also represented in historical Paintings placed under the other; all *columnation*, and the whole adorned with illuminated Ornaments'.

made out of an Oak that lately grew in St. James's Park, but was destroyed by the D[uke] of M[arlborough]. for the Great House he was building near St. James's, and that the said Oak came from an Acorn that was planted there by King Charles II being one of those Acorns that he had gathered in the Royal Oak, where he was forced to Shelter himself from the Fury of the Rebells after the Fight at Worcester. Mr. Mollineux was at the other End of the Room when this was shew'd, and the said Story told; but hearing it he comes immediately to the Table, and expresses himself in Words of this Kind, viz. that 'twas a Bawble, and that an hundred such Things were not worth seeing, Mr. Keil however thought otherwise,[165]

With no going back, the cantankerous Hearne then shows Molyneux a portrait of James Francis Edward Stuart, referred to as the Pretender (known today as the 'Old Pretender'), the episode being recorded in his diary:

I produced the Picture of a beautifull young Man, over 30 the Head of which was ΕΙΚΩΝ ΒΑΣΙΛΙΚΗ, and underneath *Quid quaritis ultra*. I did not tell them whose Picture it was, but said that I shew'd it them as a thing excellently well done, which they all allow'd, and viewed it over and over, and seemed to be mightily taken with it, and Mr. Mollineux in particular was pleased to say that 'twas admirably well done, and deserved a Place amongst the most exquisite Performances of this kind, at the same time asking how long I had had it, and whose Picture I took it to be to the former of wch Questions I reply'd, about a Quarter of a Year, to the latter that I did not pretend to tell who it was designed for. Yet Mr. Keil was pleased to laugh and to tell Mr. Mollineux they are all Rebells in this Place, speaking these Words in a merry, joking way, and not with any Intent to do me an Injury. Mr. Mollineux took the Words upon the Picture down, wch I did not deny him, not thinking that 'twas with a Design to inform against me, as it afterwards proved. Yet from this time I began a little to suspect his Integrity, and that he was not one of those good Men I expected from Mr. Keil, whom I had always found to be a Man of Honesty. I went on with my Story, and when I had done took my leave of Mr. Mollineux, who shew'd not the least Resentment, but parted with many Thanks for the Civilities I had shew'd him, and I was as glad that I had had the Happiness to oblige one that appeared to have a Love for Antiquities, and Learning, & to have an inquisitive Genius after things of Curiosity. Some of the last Words he spoke, were these, viz. Mr. Keil we will wait upon Dr. Charlett in the Afternoon, it being now after eleven, & our time too far spent for the Forenoon.[166]

165. Rannie, *Hearne*, p. 111.
166. Ibid., p. 111.

The actual picture seen by Molyneux cannot be identified today in the Oxford collections, although a small portrait on copper attributed to the French artist Alexis Simon Belle (1674–1734) was bequeathed to the university (now in the Bodleian Library) by Dr Richard Rawlinson in 1755. Belle, who became painter to the exiled king in 1701, produced numerous portraits of his patron, one example engraved by P. le-Roi carries the inscription 'the image of the king'.[167] Being such a new image of the Pretender, it is plausible that Molyneux does not recognise the sitter and questions it only because of the inscription at the head of the picture, 'ΕΙΚΩΝ ΒΑΣΙΛΙΚΗ' ('the image of the King'). It is also highly likely that Hearne is purposely goading his visitor, perhaps as a result of his perception of the Irishman's spurious knowledge and objectionable Whiggish tendencies. For Molyneux, such barefaced allegiance to the Pretender could be interpreted as treachery, thereby in his view placing the establishment at risk. Hearne records the consequence of this incident in his diary:

> The next Day after Mr. Mollineux had been at ye Library [...] Mr. Clements's Son (bookseller) told me Mr. Mollineux had a bitter Complaint against me for Shewing the Pretender's Picture, and just at this time, Mr. Keil going by the Shop, and spying me, he steps back, and says, God you will be hanged, this Mollineux will knock you on the Head, for Shewing the Pretender's Picture.[168]

Molyneux's complaint about Hearne's 'great offence' is made to Arthur Charlett, Master of University College. Hearne continues:

> After a little time he [Dr Charlett] takes me into the Gallery, and told me that he had had a Complaint made to him against me for shewing a Picture of the Prince of Wales, and another of Benj. Hoadley. I told him, I suppose this Complaint must come from Mr. Mollineux originally; which he did not seem to deny. I told him yet I did not say that one of these Pictures was the Prince of Wales, and the other B.Hoadley; on the contrary that I gave no Name to either, but left every one to judge for himself. Ay but, says he, on one there is ΕΙΚΩΝ ΒΑΣΙΛΙΚΗ,. *Quid quaritis ultra*. This I acknowledged, but added that it might be the King of Spain, or the King of Sweden, or any other King or Prince. Yet granting that I had said 'twas the Prince of Wales, I could not see the Hurt in it, since we have ye Pictures & Medals of so many other Pretenders and usurpers, &c. and yet no Offence taken at them. He did not seem however satisfyed, but begged of me for ye future not to shew either of these Pictures, wch I promised him I would not.[169]

167. Illustrated in Richard Sharp, *The Engraved Record of the Jacobite Movement* (Aldershot, 1996), p. 94, no. 132.
168. Rannie, *Hearne*, p. 113.
169. Ibid., pp. 113–14.

In a vain attempt to beat Molyneux personally in this senseless battle for scholarly superiority, Hearne's diary records disparaging remarks about his perception of Molyneux's intellect and politics.

> Nor indeed did I perceive 'till after some time that Mr. Molineux was a Man of those very bad Principles, and that debauched Understanding, as I have since found him by experience to be. Upon Mr. Keil's Request, I shew'd him some of the most considerable Curiosities in the Library. He talk'd much both about MSS. and Coyns, and, by his Discourse, one would have thought that he had spent much of his time in these Studies [...] For coming to our Cabinet of Coyns in the Gallery [...] he did not care to see modern Coyns and Medals, that would be needless for me to pull out any of those, his Genius being for Antiquities, and therefore that, if I would gratify his Curiosity, I should shew him some scarce Coyns of the Ancients. I told him we had many of those which were truly scarce and valuable, as he might soon perceive if he would cast his Eye over them. I produced two Drawers in wch are several very extraordinary ones, and wch an Antiquary might immediately judge to be good and of very great use in explaining and illustrating History. But Mr. Mollineux was so far from understanding this, that he did not discover which of them was scarce, nor what considerable Part of History might be explained from any one of them. On the contrary instead of examining these Coyns, which are genuine, and have Variety of Figures on them, he proceeded to ask whether we had a Brass Otho, a Pertinax, & other Coyns of that nature, such I mean as are look'd upon by good Antiquaries as spurious, I told him we had these Coyns, but that they had been removed as Counterfeits, and therefore not fit to be placed in this Cabinet amongst the Authentick ones. Here he spoke something about a Brass Otho in my [Lord] Pembroke's Study, which he said was certainly genuine. This made me increase my Opinion of his Confidence; however I said no more than this, that, with Submission, I much questioned it. But granting this, and other Coyns of this Kind that are sought after, to be genuine, yet me thinks Mr. Mollineux should not have inquired so strictly after these, and neglected to look upon the other Coyns I shew'd him, from wch so much Light may be drawn for ancient History. Questions about such Coyns are put by most People, I mean such as have no Judgment nor Skill in Coyns. It goes from one to the other that an Otho in Brass is scarce, and that Pertinax and others are so too. Hence when they come where Collections are they presently ask (on purpose to make People think they have some Skill) whether there be any Brass Otho amongst them, &c. This I have experienced in many Instances; and I was sorry to find Mr. Mollineux one of them, and that he was a Person tho' of good ready Discourse, yet of small Judgment in Matters of Antiquity, proceeding from taking up his Accounts from

> Conversation with Gentlemen, and not from Study, wch is a Fashion too much practised in this Age.[170]

Molyneux does not seem to dwell on this unhappy experience and immediately departs for the Divinity School (with no mention of Duke Humfrey's Library, the Convocation House and Proscholium) and the seventeenth-century School's quadrangle where he gives good descriptive accounts of the pictures and treasures.[171] Of the pictures hanging in the gallery all, except for the portraits of Christopher Columbus, can be identified in the most recent catalogue of pictures (see footnotes to the transcribed letter).

Molyneux's tame descriptions of the treasures of the Bodleian Library collections tend to confirm Hearne's opinion of his academic pretentiousness, yet his account of the Ashmolean Museum, where he presents his reader with a show of erudition based on his passion and knowledge for natural philosophy, is far from pedestrian. The museum collections were put together by Elias Ashmole (1617–92), the obsessive collector who inherited the Lambeth-based museum known as John Tradescant's 'Ark', the first popular English collection of rarities. In 1677, after a prolonged legal battle over ownership with Tradescant's widow Hester (which eventually drove her to suicide), Ashmole offered the 'Ark' along with his own extensive collection to Oxford University, which in turn commissioned Thomas Wood (*c.* 1644–65), perhaps in association with Wren's advice, to design a new building situated alongside the Sheldonian Theatre in Broad Street to house the collection (now the Museum of the History of Science) (Fig. 24).

The museum opened in 1683 and by chance Thomas Molyneux was one of its earliest recorded visitors. Writing to his brother William (Samuel's father) on 17 July 1683, Thomas tells him that

> [Dr Plot was] very civil and obliging, and shewed me all the new building [Ashmolean Museum] on the west side of the theatre, built of square freestone, containing as the inscription over the door intimates, the *Museum Ashmoleanum*; *Schola Historiae Naturalis et Officina chymica*. It consists only of these three rooms, one a-top of the other, and a large staircase. The *Museum Ashmoleanum* is the highest; the walls of which are all hung around with John Tradescant's rarities, and several others of Mr Ashmole's own gathering, his picture hangs up at one end of the room, with a curious carved frame around it, of Gibbons work [by John Riley (1646–1691). The frame by Grinling Gibbons was topped by Ashmole's

170. Ibid., pp. 109–10.
171. See illustrations by David Loggan showing 'the inside of the Bodleian Library in Oxford from ye east' and 'the inside of the Bodleian Library in Oxford from ye west' in *Oxonia Illustrata* (Oxford, 1675).

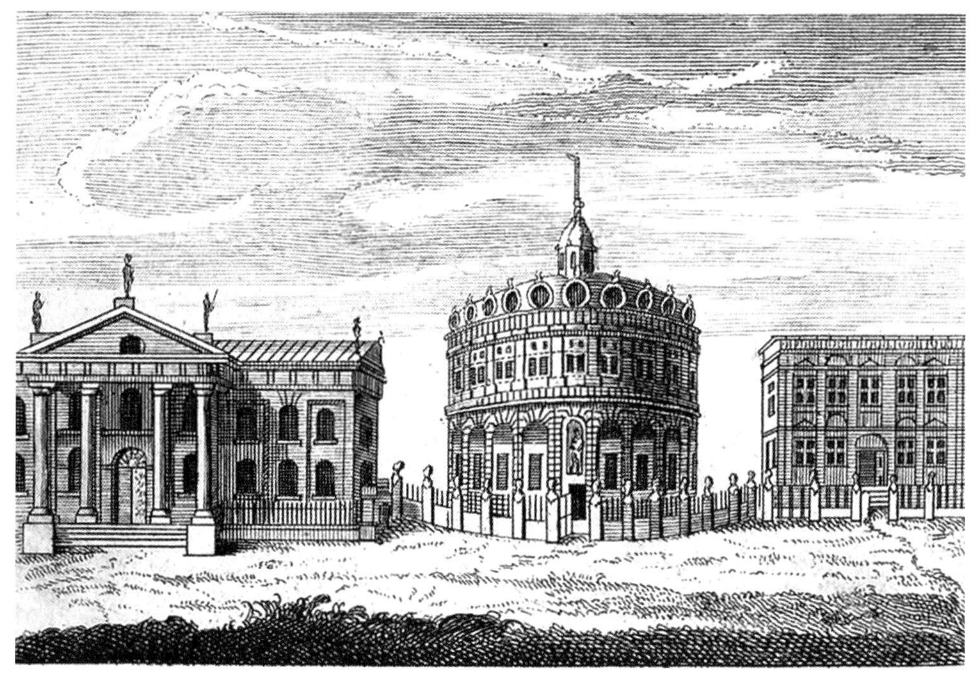

Fig. 24. Broad Street, Oxford showing Clarendon House, the Sheldonian Theatre and the Ashmolean Museum, *c.* 1720.
© *The Trustees of the British Museum*

> motto *Ex Uno Omnia* ('all things come from one'). ...] Under this room is the *Schola Hist. Nat*, very spacious and high, curiously wainscoted, at the end a very pretty white marble chimney-piece stained up and down with red [...] In this place Dr Plot reads lectures to all that go thro' a course of chymistry with him [...] Under this is the *Officina chymica*, the greatest part of which is underground, and therefore it is very cold, even in the summertime [...] it is very well contrived with great variety of furnaces, and those very convenient for all the operations in chymistry.[172]

Other commentators painted a mixed picture. Anthony Wood wrote of 'Those Doctors and Masters that pleased retired to ye museaeum which is ye upper room, where they viewed from one to 5 of ye clock what they pleased. Many that are delighted with the new Phil[osophy] are taken with them; but some for ye old — look upon them as ba[u]bles'.[173] Ralph Thoresby described the collection as 'absolutely the best collection of such

172. Quoted in Arthur MacGregor, *Tradescant's Rarities: Essays on the Founding of the Ashmolean Museum, 1683* (Oxford, 1983), p. 298.
173. A. Wood, *The History and Antiquities of the University of Oxford in two books by Anthony Wood of Merton College* (Oxford, 1796).

rarities that I ever beheld', although Edward Lhuyd in a letter to John Aubrey in 1686 expressed the view that 'the generality of the people of Oxford doe not yet know what ye museum is; for they call ye whole Buylding ye Labradory or Knackatory, & distinguish no further'.[174] Molyneux's description clearly portrays the seventeenth-century style of collecting, when natural specimens and man-made antiquities shared the same museum space. It was also a period when handling of artefacts was commonplace, prior to the trend towards the visual display of exhibits in purpose-built cabinets during the nineteenth century.[175] Three years before Molyneux's visit to the Ashmolean Museum, von Uffenbach wrote, 'it is surprising that things can be preserved even as well as they are, since the people impetuously handle everything in the usual English fashion and even the women are allowed up here for sixpence; they run here and there, grabbing at everything and taking no rebuff'.[176]

Molyneux does not specially mention handling the collections, but he is genuinely passionate and animated over Ashmole's 'very exact Catalogue' (f. **127**), *Musaeum Tradescantianum*, which had been published in 1656.[177] To Molyneux, only order and fact without discrimination could provide true knowledge and thereby reveal new possibilities by using the power of the objects to challenge the big antiquarian, natural or scientific topics of the day. Most of the artefacts he lists in his letter, including the portraits of Elias Ashmole, John Tradescant the Elder — attributed to Emanuel de Critz (1608–65), and John Tradescant the Younger — attributed to Thomas de Critz (1607–53), are instantly recognisable in the Ashmolean collections today.[178] However, sharply indicative of Molyneux's interest in nature is his interesting, albeit ambitious, comment that animals of the world could be collected in the same way as paintings or sculpture. Perhaps made with his first-hand knowledge of Sir Hans Sloane's collections in London in mind, the suggestion comes at a time when the Royal Society had already

174. Hunter, *Ralph Thoresby*; Bodleian Library MS Aubrey 12, f. 240.
175. Constance Classen, 'Museum Manners: The Sensory Life of the Early Museums', *Journal of Social History*, 40, no. 4 (Summer 2007), pp. 895–914.
176. W. H. Quarrell and W. J. C. Quarrell (eds.), *Zacharias Conrad von Uffenbach, Oxford in 1710* (Oxford, 1928), p. 26.
177. Elias Ashmole, *Museum Tradescantianum or collection of rarities preserved at South Lambeth near London by John Tredescant* (1656). For the early catalogues, see A. MacGregor, *Ashmolean Museum, Oxford. Manuscript Catalogues of the Early Museum Collections 1683–1886 (Part 1)* (Oxford, 2000). A consolidated copy was prepared for security purposes in 1695: see A. MacGregor and M. Hook, *Ashmolean Museum, Oxford. Manuscript Catalogues of the Early Museum Collections, 1683–1886 (Part 2)* (Oxford, 2006).
178. The elder Tradescant was gardener to the Earl of Salisbury, Lord Wotton, the Duke of Buckingham and King Charles I, while the younger succeeded his father as Keeper of His Majesty's gardens, vines and silkworms at Oatlands Palace.

employed a collector to assemble specimens of native animals to Britain. Such thoughts would have stimulated the relationship between Molyneux the antiquarian and Molyneux the natural historian, where the process of enquiry was essentially the same. On a par with such enquiry is the interaction with science, so it is appropriate that Molyneux refers next to the Natural Philosophy School and to the chemical laboratory in the basement of the Ashmolean Museum, with its thick barrel-vaulted ceiling that protected the rest of the property in the event of an explosion.[179]

During his visit Molyneux meets the museum's keeper, David Parry (*c.* 1682–1714), who in conversation gives an account of his Irish travels from May 1697 with the Welshman, Edward Lhuyd (1660–1709), one of the greatest antiquaries and naturalists of his day.[180] In 1689 Lhuyd was appointed as assistant to Dr Plot, Keeper of the Ashmolean Museum, and succeeded him in 1690. Thomas Hearne reckoned Lhuyd to be 'naturally addicted to ye Study of Plants, Stones, &c as also Antiquities'. His catalogue of British fossils, *Lithophylacii Britannici Ichnographica*, was published in 1699 in an edition running to only 120 copies (a second revised edition was not to appear until 1760). Molyneux indicates that he already knew of Lhuyd, though whether he had met him personally or had only corresponded we cannot be sure, and reveals that before his unexpected death Lhuyd had promised him a transcript of his journal made in Southern Ireland whilst gathering research for his *Archaeologia Britannica* (1707). Reference is also made to a dispute that arose over ownership of Lhuyd's collection of minerals which was never satisfactorily resolved. Parry, Lhuyd's unpaid assistant until 1709 and successor as Keeper, appears to have been helpful to the Irishman despite many contemporaries claiming that his weakness for the Oxford taverns made him an ineffectual curator.

Although Molyneux is pleased on the whole with his visit to Oxford, he does express some disappointments. His confrontation with Hearne is compounded further by his disapproval of fees being levied for access to colleges and collections (a point also remarked on by von Uffenbach).[181] Moreover, Molyneux shows some apprehension for the welfare of the marbles displayed in the external niches of the Sheldonian theatre. Donated to the university in 1677 by Henry Howard, the Duke of Norfolk, the marbles had been collected by Sir Thomas Howard, the 2nd Earl of Arundel (1586–1646) and had been on display at his home, Arundel House in the Strand in London, prior to its demolition in 1678. Molyneux expresses

179. Refer to Anthony Simcock, *The Ashmolean Museum and Oxford Science 1683–1983* (Oxford, 1984).

180. Arthur MacGregor, 'Edward Lhuyd, Museum Keeper', *Welsh History Review*, 25, no. 1 (2010), pp. 50–73.

181. Quarrell and Quarrell, *von Uffenbach in Oxford*, p. 17.

justifiable concern that the intrinsic value of the marbles would be lost if the carved lettering became 'defaced' through exposure to the elements, though he takes some solace from the fact that the inscriptions had been recorded and published in *Marmora Oxoniensa* (1676), by Humphrey Prideaux (1648–1724). The marbles are now safely housed in the collections of the Ashmolean Museum.[182]

Nor does Molyneux approve much of Oxford's architecture. Sir Christopher Wren's theatre, built between 1664 and 1669 at the expense of Gilbert Sheldon, Archbishop of Canterbury and former Warden of All Souls College, receives an enthusiastic account of its interior which, says Molyneux, '[...] must compensate for the Transgressions of its outside' (f. **132**). Again, he appears weighed down by his dislike of the Baroque, a viewpoint that was typical of the architectural debates of the day. Mention is made of the published engravings by David Loggan (1634–92) of the Sheldonian Theatre in *Oxonia illustrata* (1675) and, as with his description of the prospect of St Paul's Cathedral (f. **35**), Molyneux shares the concerns of James Yonge who wrote in 1681 that the Sheldonian Theatre was 'pestered up with the too near being of other houses'.[183] More positively, he praises the function of the Oxford University printing house, then situated beneath the Sheldonian Theatre but in the process of being relocated to the Clarendon Building, newly erected by Hawksmoor between 1711 and 1713 and home of the University Press until 1830. Of this handsome Palladian structure Molyneux says nothing other than it was 'noble' (f. **133**).

On Sunday 22 February, it is likely that Molyneux attended church at the 'Gothick' (f. **134**) University Church of St Mary the Virgin before braving bad weather and appalling roads on his journey to Blenheim Palace (Fig. 25). Some distance beyond Blenheim Molyneux considers, but does not visit, Heythrop House, built between 1707 and 1710 by the amateur architect Thomas Archer for Charles Talbot, the 12th Earl and Duke of Shrewsbury. Although the designs of the two houses are similar, Heythrop, like other houses by Archer, was influenced by the continental Baroque of Bernini rather than that popularised by Vanbrugh.[184] It may well be that he had every intention of seeing Heythrop but was beaten by the weather, hence his reader has to make do with illustrations which Molyneux sends along with plans of Blenheim Palace.

Molyneux's account of Blenheim, although not the earliest, is certainly one of the most complete of the building and landscape still under

182. Refer to Michael Vickers, *Arundel and Pomfret Marbles in Oxford* (Oxford: Ashmolean Museum, 2007).
183. F. N. L. Poynter (ed.), *The Journal of James Yonge 1647–1721* (London, 1963), p. 179.
184. After a fire in 1831, the house was left as a ruin until bought by the wealthy railway contractor, Thomas Brassey, who commissioned the architect Alfred Waterhouse to refurbish the house. Today it is a hotel and conference centre.

Fig. 25. Blenheim Palace, a north-westerly view showing the Grand Bridge, and Blenheim Park with the village of Woodstock in the distance. Eighteenth-century English School.

© *NTPL/Derrick E. Witty*

Fig. 26. John Churchill, 1st Duke of Marlborough (1650–1722).

construction. Built on the site of the palace and deer park of Woodstock, on land given by Queen Anne with monies awarded by Parliament, the Baroque palace was designed by the architect John Vanbrugh and named after the place where John Churchill, the 1st Duke of Marlborough (Fig. 26), was victorious in battle over the French and Bavarians on the River Danube in 1704. Work had, however, stopped in December 1711 when Marlborough fell from favour and was dismissed by the Tory ministry. Despite his personal regard for the duke, Molyneux cannot conceal his disappointment in Blenheim. Rather than being a symbol of nationhood and identity or, as he put it, 'the Seat of Valour and Beauty' (f. **139**), he considers the incomplete building as

> little an Honour to the Nation as to the Builder that undertook it and I must confess I could not but agree with what a disaffected gentleman of our company say'd pleasantly enough of Blenheim tho I did by no means give into his insinuated Application that for his part take it all together, it had the very Air of a heavy Weight & Oppression on the Earth that bore it. (f. **139**)

Despite the accuracy of his account, there is a tendency for Molyneux to present an over-romantic image influenced, it would seem, by his unidentified guide. As such, he goes into some detail about King Alfred's associations with Woodstock Park as well as recounting stories of King Henry I and his menagerie of exotic wild species, and of King Henry II courting the fair 'Mistress Rosamond [Clifford]' (f. **140**), who was eventually poisoned by Queen Eleanor (the subject of an Addison and Clayton opera of 1707). Furthermore, despite the 'unhappy season' and 'worse weather' (f. **135**) he claims to have seen three or four churches from the south-facing 'garden door', which is not possible in reality, and implies that the unfinished building is more complete than it actually was. There was at the time a network of antiquarian groups with an interest in saving ancient ruins, but Molyneux refers only in passing to the ruins of Woodstock Manor, merely writing 'the Walls [...] do partly remain' (f. **140**). His description offers no melancholy delight in the remains, nor does he seem to appreciate their aesthetic or picturesque worth, a viewpoint shared by the Duchess of Marlborough but not by her architect Vanbrugh, who was very keen to salvage the old manor for antiquarian effect. The ruins that Molyneux describes were the result of moving some building rubble from the old manor house for use in the Grand Bridge, a procedure that was recorded as early as 1708, while the ruined house itself was illustrated by an unknown artist in 1714.[185]

Not surprisingly, Molyneux is disaffected by the overtly military and threatening nature of Blenheim's Baroque architecture, considering the portico 'Clumsy and Crowded', the 'several Towers [...] extreamly Gothick and Superfluous', the north entrance front 'surprizeingly odd' and the statues 'clumsy and ugly' (f. **135**). His judgment is echoed by *The Spectator*, which described the pediment depicting 'the Figure of a monsterous Lion tearing to Pieces a little Cock [...] lately hewn out of Free-stone, and erected over two of the Portals of *Blenheim* House' as 'Such a Device in so noble a Pile of Building looks like a Pun in an Heroick Poem; and I am very sorry for the truly ingenious Architect would suffer the Statuary to blemish his excellent Plan with so poor Conciet'.[186]

The less menacing south front is more agreeable to Molyneux's sensibilities, though not for the 'Ugly Towers you see it loaded with' (f. **139**). The Grand Bridge, then marooned in an unfinished landscape, is also to Molyneux's taste, perhaps a consequence of its streamlined Palladian classicism which was the hallmark of the style of Nicholas Hawksmoor rather than that of the Baroque Vanbrugh. However, he fails to mention that only the western side of the bridge had been finished (the arch and two western towers being completed in the autumn of 1711) or that work had been abandoned altogether in 1712. Sadly, he does not describe the bridge's interior rooms (of which no written record survives) but does write, 'The

185. Hunt and Willis, *Genius of Place*, p. 199.
186. *The Spectator*, 8 May 1711, p. 231.

Arch I believe is at least 40 feet high and has in its Remparts and Buttresses all the Conveniencys of Water Engines to supply the house, bathing rooms and the like' (f. **136**). A few years later, Stephen Switzer mentioned the 'Water Engine under the Small North Arch of Blenheim Bridge' which was certainly working by 1706, and in 1728 Fougereau, a Frenchman visiting Blenheim, wrote of the 'fresh water baths'.[187]

Molyneux's account of his tour of the palace chambers is remarkably accurate. On entering at the north entrance he passes through the Hall, the Salon and the nine chambers on the south garden façade which adjoins the east and west wings. In the cellars Molyneux spies a fountain, depicting the gods of four great rivers from four great continents, modelled from the 1651 original by Gianlorenzo Bernini situated in the Piazza Navona in Rome. The copy was presented to the Duke of Marlborough by the Spanish Ambassador to the Vatican but was not installed in the gardens until 1774 (now repositioned in the lower water terrace garden).[188] Of the undeveloped gardens, established by Henry Wise but by 1713 largely in the care of Charles Bridgeman, Molyneux notes that they 'will be beautiful' (f. **138**). William Stukeley wrote of the Blenheim gardens in 1712:

> The garden is [...] taken out of the park, and may still be said to be part of it, well contriv'd by sinking the outer wall into a foss, to give one a view quite round and take off the odious appearance of confinement and limitation to the eye.[189]

Molyneux misses this specific point, but the concept of returning a garden to nature by connecting with the landscape clearly meets with his approval.

From Blenheim, Molyneux visits Stonesfield near Woodstock, where he tries unsuccessfully to see the Roman pavement which had been discovered on 25 January 1712 by a farmer, George Handes or Hannes, when 'his Plow-share happen'd to hit upon some Foundation-Stones, amongst which, he turn'd up an URN' and in searching further he laid bare a figured tessellated pavement measuring 35 ft (10.7 m) by 20 ft (6.1 m) 'not above 2 Foot [0.61m] under Ground'.[190] This discovery, which later incorporated a large Roman house close to Akeman Street, attracted the attention of Thomas Hearne who, as Molyneux indicates, published an account in the eighth volume of *The Itinerary of John Leland the Antiquary* (1712) along with an engraving

187. Stephen Switzer, *Introduction to a System of Hydrostaticks and Hydraulics*, II (London, 1729), p. 321. For Fougereau's visit to Blenheim, see Howard Colvin, 'The Grand Bridge at Blenheim', in *Essays on English Architectural History* (New Haven and London, 1999), p. 258.
188. D. Green, 'The Bernini Fountain at Blenheim', *Country Life*, 27 July 1951, pp. 268–9.
189. William Stukeley, *Itinerarium curiosum* (London, 1724), Itinerary II, p. 44.
190. J. Pointer, *An Account of a Roman pavement lately found at Stunsfield in Oxfordshire* (Oxford, 1713). For a history of the site, refer to M. V. Taylor, 'The Roman Tessellated Pavement at Stonefield, Oxon', *Oxoniensia*, VI (1941), pp. 1–10.

by Michael Burghers.[191] The pavement has not been uncovered since 1973. Molyneux then returns to Oxford before departing on the Monday 23 February for an overnight stay at Abingdon. On 24 February he travels to Henley, the following day to Maidenhead and then arrives back in London on Thursday 26 February, passing the seats of Sir William Compton, Mr Knight and Lord Orkney's house Cliveden, recently restored by the architect Thomas Archer for Lord George Hamilton, Earl of Orkney, a senior general in the Duke of Marlborough's army. Hence, as Molyneux declares, his journey has been completed 'exactly in a Week' (f. **142**), between Thursday 19 and 26 February 1713.

Postscript (ff. 142–6)

The postscript describes some London-based collections associated with Sir John Germain (1650–1718), Thomas Pitt (1653–1726), James Petiver (1658–1718) and a 'Mr. Roberts'. Although undated, the visits coincide with his return to the city, so therefore took place between 26 and 28 February, when this letter was written.

Well known to the 1st Duke of Marlborough, Sir John Germain (Map I ⑲) inherited the Earl of Arundel (Thomas Howard)'s notable gem collection from his first wife, Lady Mary. Molyneux mentions that the gems had been catalogued, which may refer to one of Arundel's own catalogues or to the early eighteenth-century catalogue compiled by Sir Andrew Fountaine, later published in 1731.[192] The suggestion that the gems are being offered for sale on the open market when Molyneux has sight of them is unsupported, as are his concerns about the sale of the collections, which did not happen until 1875 when they were auctioned by the 7th Duke of Marlborough.[193] Molyneux views the gems, most likely at Germain's house in St James's Square, in the company of 'our good friend' (f. **143**) Lord Pembroke. In conversation Pembroke speaks of the Pitt diamond, purchased by Thomas Pitt, lately Governor of Madras, for the 'bargain' price of £53,000. It was later sold to the Regent of France for £135,000, in order to raise capital for the purchase of the Boconnoc estate in Cornwall, and was

191. Thomas Hearne, 'A Discourse concerning the Stunsfield tessellated Pavement', published in his edition of *The Itinerary of John Leland the Antiquary*, VIII (Oxford, 1712), Preface, pp. iv–xxxii. Cosh, S.R. and Neal, D.S., *Roman Mosaics of Britain* (London, 2010) Vol. IV, p. 212.

192. Andrew Fountaine, *The Arundel Cabinet* (London, 1731). For the latest research on the gems, refer to John Boardman with Diana Scarisbrick, *The Marlborough Gems* (Oxford, 2009).

193. The collection was acquired from Sir John Germain's second wife in 1769. It was later catalogued by Jacob Bryant (translated into French by L. Dutens), *Gemmarum antiquarum delectus; ex prae stantioribus desumptus, quae in dactyliothecis ducis Marlburiensis conservantur ... Choix de pierres antiques gravees, du cabinet du duc de Marlborough*, 2 vols (private, 1781).

reputedly installed into Napoleon's sword before passing into the collection of the Louvre in Paris.

Rarely in his letters does Molyneux display a social conscience, yet in this postscript he writes approvingly of the charitable status and benevolent function of the house of correction, known as Bridewell (Map V ⑳), in Tothill Fields, Westminster. Much improved during Queen Anne's reign, Bridewell became an exemplar of social correction through 'the setting of idle and lew'd people to work', a system greatly applauded in von Uffenbach's account of 1710.[194] The site of Bridewell is now taken up by Westminster Cathedral.

Undoubtedly the highlight of his return to the city is his visit to James Petiver's museum (Map III ㉑) of botany and entomology housed in his apothecary shop in Aldersgate Street. Petiver (*c.* 1665–1718) had been elected as a Fellow of the Royal Society in 1695 and was regarded by the naturalist John Ray (1627–1705) as 'best skilled in oriental and indeed in all exotic plants of any man I know', although von Uffenbach thought him 'wretched in both looks and actions' and Sir Hans Sloane was concerned over his disorganisation.[195] In 1718 Sloane bought Petiver's museum to add to his own extensive collection. Regardless, Petiver's natural history collection was the best of his generation and his published guidance for collecting specimens was regarded as the standard text on the subject.[196] Molyneux's narrative highlights the global wealth of the collection, a philosophy much advocated by the Royal Society, but, like Sloane, directly criticises his host's lack of formal education and indirectly his unsystematic way of working.

The identity of 'Mr Roberts' (f. **147**) is uncertain but, judging by the quality of the timepiece described, Molyneux is probably referring to Francis Robartes (1649–1718), uncle of the Whig Member of Parliament, Charles Bodville Robartes (1660–1723), the 2nd Earl of Radnor, who by February 1713 had fallen on hard times. Francis, a younger son of John Robartes (1606–85), the 1st Earl of Radnor and Lord Lieutenant of Ireland between 1669 and 1670, had strong links with Ireland, serving as a privy councillor and revenue commissioner under King William III between 1691 and 1692, and was elected to the Irish parliament in 1692 and became president of the Dublin Philosophical Society in 1693. He was elected vice-president of the

194. S. and B. Webb, *English Poor Law History. Part 1 — The Old Poor Law* (London, 1963), p. 50; Quarrell and Mare, *von Uffenbach*, pp. 53–5.
195. Quoted in Arthur MacGregor (ed.), *Sir Hans Sloane, Collector, Scientist, Antiquary* (London, 1994), p. 23; Quarrell and Mare, *von Uffenbach*, p. 127.
196. James Petiver, 'Brief Directions for the Easie Making and Preserving collections of all natural Curiosities', in *Opera, historiam naturalem spectantia; or gazophyl. Containing several 1000 figures of birds, beasts, reptiles, insects, fish, beetles, moths, flies, shells, corals, fossils, minerals, stones, funguses, mosses, herbs, plants, &c from all nations, on 156 copperplates* (London, 1767).

Royal Society in 1704. The clock, most likely seen at Robartes' home in Pall Mall, appears to be an earlier version of a Congreve rolling ball clock, first described in Gaspar Schott's *Technica Curiosa* (Würzburg, 1664); a model well known by Molyneux's time but here very loosely explained.

Text of Letter 5 in two fragments

[**105**] Dear Sir Febry 28.1712/3 London

On Thursday the 19th Febry I took a progress in order to see Oxford; we were in the Coach by nine a clock and I think in about 4h ½ we arriv'd thro' a very fine improv'd Country at Windsor, Wee left Hampton Court on Our left and nearer Windsor a house of the Duke of Montagues which we had not time to see upon Our Right; Windsor you see is situated in Berkshire on the Thames, the Situation is so elevated however you see it at a great distance and indeed in a clear day from London but as you come nearer it makes in my opinion a most noble Prospect and has at all distances as much the Air of a Palace as any I have Seen; it consists of a very large irregular Court of Buildings, the one side being entirely taken up in the State Apartments the others in several Offices and lodgings for the Lords Treasurer, High Steward &:c as usual, this Court is only gravell'd and not Levell and yet the buildings that enclose it being all of mighty good hewn Stone, tho' of the old Gothick Architecture and built castlewise have something so grand and uncomon as to Set off the Court very well. I must not forget to tell you, that you have here in the middle a Statue on Horseback of King Charles the 2nd. On the Stairs going up to the Apartments I observ'd a few ancient Busts, the Stairs and indeed every room I think [**106**] we saw is crowded with various paintings on the Ceilings; the apartments I think are Somewhat Irregular so that I can't so well describe them as in so old a building they generally are. I can only tell you that they have more room here than in any other of the Queen's Palaces, and I think three different Apartments with distinct Bedchambers pretty compleat; in all of these the Furniture has nothing extraordinary except the Pictures, which are in my opinion excellent and in great abundance in all the Rooms, but more particularly in a private Gallery belonging to the Queen's Apartment, There are several most incomparable heads by Titian as they assur'd us which I thought the best Pictures I ever saw in my Life, Among these I particularly admir'd the Portrait of Johannes Duns Scotus the famous Philosopher, an Original as 'tis supported, There is one History piece also of Sr Thomas Gresham buying the Famous Jewell from the Jew, which the Queen thought too dear for her and he is said to have downed and drank to her good health, which Picture I mention as well because that it has as vivid colouring as is now as full of life as possible because it was done by the famous Smith, one Quintin, who marry'd I think Titian's Daughter as the Story goes by the Strength of this Qualification Love has often made Originals but so incomparable a Copy of Nature as this is I never heard of its performing [**107**] before. They shew'd us several others of Titian, Raphael and Palma in other places and indeed I think I did not Spy one bad piece in the whole Palace, and could have hardly thought there were so many good ones together in England as I saw here. St George's Hall which is part of one of the apartments is worth seeing 'tis painted Ceiling & Sides by Varrio, the Sides bearing the History of Edward the 3^{rds} triumphs in France who you know was Stil'd the Black Prince, and was Institutor of the most noble Order of the Garter. The Subject gave me more Pleasure than the Execution and yet to tell you the truth I could not but wish the like Accident may never happen again and am willing to compound with Fortune if Brittain & France shall never again know the same King. At the end of this Hall you see the late King painted on a Throne rais'd on Steps which indeed are perfectly well done when view'd from the true point of Perspective in a Gallery at the other end of the Hall; adjacent to this Hall is the Pallace Chappell and here which is pretty enough you have from the Comunion table of the one to the far end of the other which I believe may be about 200 feet a thorough Vista, this also is all painted as the former and by the same hand; in one of the Guard rooms adjacent to this Hall they were raiseing Scaffolds and Gallerys around for the ensueing Installment of several Knights of the Garter which they suddenly expect. I must not forget to tell you that in [**108**] one of the Queen's Closetts we observ'd several little Pikes with Silver heads and Sattin Flaggs with Fleurs de Lis on them not above 3 feet long; we enquire'd what they were & it seems these are the Duke of Marlborough's Rent for Woodstock Mannor. Having seen the Apartments we went to the Constables Lodgeings now the Duke of Northumberland's[197] which are in a large round stone Castle on a vast high Mount but within the [gap] of the Pallace surrounded with a deep Trench and made very impregnable against Bows and Arrows and was I suppose design'd as a kind of Cittadel to the Court which was certainly built as a Fortification & place of Strength as well as a Pallace, On the Leads of this Tower we observ'd laying several Tin Chests of Powder. The East and South Sides of this haughty Stone Pallace and part of the West side have round them below a noble large Terras=walk just on the Brow of the Hill on which it stands the wall of which Terras is all of Stone hewn and was part of the Strength of this place being above 20, or, 30 Feet high at least. Below this especially on the East Side, which is the longest Front I believe at least 400, or, 500 yards long the Hill which is extreamly steep is cut into 3 or 4 very noble regular Slopes, and after that you have a vast large & noble plain perhaps of 100 acres or thereabouts design'd by King William to be made into a Noble Partere to be seen from [**109**] the commanding height of the Terras, and beyond that again the River

197. George Fitzroy (1665–1716), Duke of Northumberland, Constable of Windsor Castle and ranger of Windsor Forest between 1701 and 1714.

Thames, and here I must confess that if this design were executed as I pleas'd my self with imagining it the Exalted & noble Situation of this Pallace I believe at least 105 or more yards Perpendicular from the River, the unbounded Prospect it commands & the Beautifully wooded Country you have in view, the numerous dispers'd villages & the near one of Windsor just as it were under you together with the house Park which is close to the South Terras and for some way nearly of a level with it, would in my mind compose as happy a juncture of Circumstances for a great and noble Pallace as I am capable of conceiveing, From the Terras we walked a little into the Park to take a View of the Pallace from thence and to tell you the Truth, tho' we saw but the worst & Shortest the East Front, I must own its great height & Strength and the aspireing Magnificence it borrows from its Situation exceedingly struck me with Pleasure and Surprize And plainly convinces me of what I have often fancied, that if Our Forefathers had not so much of Art as we have they had at least as much of Grandeur & use and indeed in Windsor there is no superfluous Gothick Ornaments, but something of that happy sublime Naivete & Greatness in its Walls & Situation something so unconfinedly regular as to tell you the Truth I never yet saw in any of the modern buildings. It seems to partake [**110**] of the Taste of its first Builder who was I think W[m] the Conqueror & methought whilst I survey'd its lofty Towers I fancy'd it could not have been built but by the same dareing Genius that furnish'd it with a Waste of 20 Miles round as a Forest and Scene that he could view, in short in all respects I must confess Windsor is the only Pallace that comes near the Ideas of Pomp and Superiority I have long given in my mind to the Dwelling of a Prince. How easy does Nature here give surprizeing pomp and Beauty which the utmost Art in other places where she denys it cannot attain. S[t] George's Chappel and several lodges of the poor Knights of Windsor with the Guard houses are seen in the outward Courts, the first is only worth viewing being the Place where the Knights of the Garter are install'd, 'tis an old Gothick building with the Choir exactly of an old Cathedrall and Stalls at each side where the Knights and Prebends sit alternately mixt, in the Knights Stalls you see the Arms of all the Knights that have ever been of this order painted and fixt up there on Plates of Brass and over head hang Banners with their Arms at present; The wainscot & carveings are extreamly Gothick and old, and indeed I such scarce have mentioned the fam'd S[t] George's Chappell but to tell you there hangs there over the Communion table a most excellent piece of painting representing our Saviour at the Passover, I enquir'd whose hand it was which [**111**] they did not know, but we fancy'd it was Raphaells. Before we leave Windsor I must tell you there is besides the Park I mention'd another vastly larger where are 2 or 3 lodges very well worth seeing as they told me, but we had not time to go there and therefore saw no more of that Park but its Prospect at a Distance, We could Pe'ceive in it a most noble long Walk of grown trees directed to the Palace and several Woods here and there dispers'd but I must tell you for your Comfort that I have given but a slight Description of Windsor and its Beautys you will soon I assure you see a very just and agreeable one in a Poem still'd Windsor Forrest which will suddenly appear. Fryday ye 20th wee took our leave of Windsor and in about 4½ hours arriv'd at a little Village in Oxfordshire called Nettlebed. I remember we pass'd this morning through a small village the seat of one Mr Pall. The Country was in many places healthy, full of Woods, Hills and Bottoms where Robberys they tell me are often committed. From one of these Hills you have a Prospect of Hirley Bottom thro' which runs the Thames, you see Hirley there a Seat of Lord Lovelaces and on the other side of the River Buckinghamshire and another seat on the very Bank very pleasantly situated of Sir James Etheridge. Wee passed over the Thames at Henley which is a pretty good Town in Oxfordshire and is said by Dr Plott [**112**] to be the most antient in this County. I must not forget to tell you that I observ'd all this morning in several places the Soyl as the same Doctor I think describes it entirely Chalk after two feet depth so that on all the Broken sides of Hills you see great quantitys of it and in many Places it has made the Roads I think almost as bad as Chalk Hill at Dunstable. I observ'd also as usual in Chalk great quantitys of flint Stones and in many places they seem to have no other for their houses and buildings. This is a very surprising appearance in nature which the Naturalist can't get over, that the hardest Substances should be found in the softest beds. Morallists perhaps may make the same observation in the genius of mankind and I think much more accountably. This Nettlebed is said to be the highest point at least of these parts of England. From hence after dinner we pass'd thro' a Country in Spotts here anhd there mightily beautiful and well improv'd in others more open & wild by Ewelme & Benson two almost contiguous and very lovely Shaded Villages, in the first of which Our Kings as it is said had formerly a Residence and soon after by Dorchester and Sandford 2 other Villages and arriv'd in 4hr½ more at the Famous City of Oxford. The greatest glory of this place you may be sure I esteem that that all the World esteems the greatest Glory of Brittain I mean the very famed and [end of folio **112**]

[Folios **113–17** missing]

[**118**] Square Pillars particularly I remark'd, for which 'tis sayd the Duke of M——— offer'd £5ld, but was refus'd they are at least 5 or 6 feet Square & I believe about 20 high. But of all their Chappells the most considerably beautiful is that of Trinity College, This is pav'd with black and white Marble, wonderfully neatly wainscotted, the Screen and Altar piece entirely of Cedar, & all together I assure you makes a very beautiful Chappell as ever I saw in my life, From hence I think we step'd into Baliol College Library, which has nothing extraordinary but that 'tis say'd to contain more of the best Editions of the Classick Authors than any other in Oxford. I think I have now told you most of the remarkeable things I saw in the

Private Colleges, and have nothing further to say on this head, than that in General they are much less than ours in Dublin for the most part consisting of not above two little Stone Quadrangles sometimes with Cloysters or arch'd walks round them, The Halls and Chappells are also proportionately small but generally adorn'd with Pictures of their Founders. I remember in one, I think University College Hall, I saw the Picture of King AElfred[198] the Saxon King which is certainly Fantastical & without any Authority at all tho' it is also sure that there is a College in Oxford and I think it is this one that was not so properly founded as restor'd by this Prince being [**119**] much more antient than his time, I observ'd I think a Brick hearth for a fireplace in the middle of every Hall and in all the Librarys I visited I generally took care to enquire whether their Manuscripts agreed with the printed Catalogue and they generally answer'd me they did, For the Paintings in their Chappells, haveing never seen what they have abroad I don't know how far they are to be like'd but for my own part I must confess I like'd them very well & though they diffus'd such a venerable gloom thro' the whole Chappell that they were very useful as well as beautifuls I am sure may on this Account be esteem'd very decent helps to the Aire and Devotion of the Place. I stay'd so short a time at Oxford that I can give very little Account of the different Genius'^s^ of each College. I can only assure you that there is only much Emulation between the whole & their Sister of Cambridge but also between each College and the rest as has I am confident had a very useful influence on the Studys and Productions of each particular Society and I can tell you besides that if they Emulate in these respects they do so no less in being civil to Strangers. However I must own I learn'd enough to know that some Colleges are particularly resorted to by some Countrys, the West of England goes to one, the North to another, the Scotch I think to Balioll & the Welsh to Jesus, the Witts and Belles Lettres flourish I think [**120**] at Christs Church & Queens is more particularly remarkeable for natural Philosophy and Physicks: The great and many Curiositys of their Publick Buildings remain yet to be spoke to and these I think we may reckon the Bodleian Library, the Anatomy School, and the other Schools, the Museum Ashmoleanum and the Theatre. These Buildings are entirely common and belong to each College as they compose one body in a University. The Bodleian Library was what I first visited, haveing a Letter to D^r:^ Hudson who is keeper of it and indeed I must confess my Surprize was very great in seeing the infinite Number of Books that are here amass'd together, The Building is in the Shape of a H very large and very high with Gallerys above clear round so that the Books are plac'd only against the Walls and by this means make a much more magnificent Figure than in those Librarys where the Shelves come out towards the Middle of the room, To determine their Number is beyond my Power, I can only tell you, in one quarter you

198. An image of King Alfred was commissioned in the 1660s and still survives today.

may see a compleat Library of one sort of Books, in another as large a one of some other Science, in a third a prodigious number of Manuscripts of all Languages, particularly the Oriental, so that I can grossly say the Bodleian Library seems to contain may be 4 or 5 times as many Books as ever I saw together in one room in my Life and all of them in the lower Shelves under **[121]** the Gallerys chain'd as in most of the other Librarys in Oxford. Wee were assur'd by Mr Hearne the under Keeper, that the Number of them is at least doubled since the Catalogue of them was printed and I believe there may be possibly 100,000 volumes or more; What infinite unconquer'd Worlds are here of Science, what Grief, what envy must it extort from the Insatiate Philosophick mind, to view the untasted Volumes that the Shelves contain, not Alexander with more mistaken pleasure sought another World, How much more prudent the humble ignorant content to think the Prize not worth the Labour and to disdain the useless Conquests, in Effect as wisely say the Scriptures, there is much Sorrow in much knowledge and it cannot be believ'd that Providence should have admitted so many Books into the world if very many of them had not been Trifles and unnecessary to be look'd into by a Wise man, I am bold to say in its defence that 3 in 4 of those that have been written have very justly been condemn'd to the Oven or to the Bodleian Library and for the peace and quiet of the World I can't but wish the proportion had yet been greater; in these Reflections possibly one may fall out with the Art of Printing and we may not think ourselves in these Ages so much oblig'd to it as we are generally believed, to confess a Truth I have always thought the Light that this Art has brought into the World was such a one as has brought a Fire along with it, and as we have less of Ignorance **[122]** so we have less of Quiet and Content since this Invention and I must confess I am sometimes tempted more to esteem the light that burning Pharos gave the Antient World; Before we leave this vast Repository of Books I must not neglect to tell you that the most curious of Manuscripts they particularly chose to shew us, were a vast large Greek Paper Folio containing a Collection of many unprinted Passages of antient Fathers by Niceta, This is one of the fairest MSS: I ever saw in my life & I think they design to print it, Manael Phile of Birds, a Greek Poem which belong'd formerly to the fam'd Printer Stephens and from the Letters of which he made his Types as it is say'd, the MSS of John Lelands Itinerary lately publish'd, Beda pretty antient on parchment, The Acts of Apostles, and certain Lessons or Excerpta from Gospells both these in Capitalls and near 1000 years old and pretty fair; They shew'd us also a more modern MSS of the Old Testament to Job done in Little Oval Pictures upon gilt grounds exactly like one of the Gospells Lord Treasurer has as I told you and possibly may be the first Volume of the same Work. From the Bodleian Library we were conducted by the same Gentleman who shew'd it us Mr Hearne, who by the way is a very learn'd Man & to whom the World is much obliged for several fine Editions; to see the Anatomy School which is under his Care & with the Library and other Publick Schools composes a good handsome square Quadrangle of Hewn Stone [**123**] which is not the least beautiful one in Oxford. In the Anatomy School you see a Museum consisting of several Natural and Artificial Bodys which are not kept in the best order nor I believe so much as Catalogued, I saw nothing there very remarkeably Curious except it were a Map of China, a Turkish Map of the World, that is done in their Language, a vast large Family Urn I believe at least 2 feet high & proportionable and which was indeed a very remarkeable and extraordinary curiosity, the right Thigh bone of a Man I think dug up in S[t] Clements Church yard in London at least 3 feet and above long and every way of a due Correspondent bigness, there hung by a Skeleton and we compar'd the two bones & were of opinion that the large one was in all respects as well shap'd and proportion'd as the other & was at least twice as long tho' the Skeleton was none of the least. Among the Artificials we could not but Smile to see the care that is taken even in these trifles to inculcate the Principles for which this University is fam'd, You see a Silly cut of Mr Hodly with horns & asses Ears and they tell you his Wife gave the present. You have a Server made of the Royal Oak Wood nay even a scurvy Tobacco stopper which with great ignominy you are sure to hear was made of the Wood of an oak that the D. of M——— h ungraciously rooted up in S[t] James[s] Park because it stood in his way when he built his house there [**124**] and this particularly for his inveterate Malice to Monarchy & so forth, for that he said Oak was planted there by King Charles 2nd and grew from an acorn that was taken from the Royal Oak, I am asham'd to mention such Follys to you, but if you had been plagued with so much of this Cant as I was wherever there could be an Occasion taken in shewing the Baubles I am sure it had not fail'd to have made some impression on you, You are not to apprehend that I believe this arises entirely from the direction of the University, I believe it arises in a great measure from the impertinence and deep political Genius of the Shewer, and this I am the more inclin'd to hope because among trifles, I do assure you the Gentleman shew'd us not without much Smiles and Laughter the Picture in Black & White of a person whom the Religion and Loyalty of this University do by no means permit them to Esteem, whether the merry manner he was pleas'd to produce it in did seem to abet and approve of the Picture or not I don't know but I think no one can shew it without approbation, and few I hope will see it without Horrour, when they see written above EIKΩN BA I IKH and below, *quid quaritis ultra*, I have heard since that this Picture was one of some thrown into the English Camp in Flanders some years ago with Medalls of the same Person and that he pretends to shew it on that Account; However I am sure the Language of the [**125**] Inscription seems to make it so much more proper a present for Scholars than for Soldiers, that I am confident none of the heads of the University know of this Publick Scandalous folly, which among the prejudic'd can't fail to do them great Dishonour. Haveing seen the Anatomy Schooll we look'd into two of the others, the Astronomy & Geometry Schools and found them all like one another only large rooms with Desks for the several Professors at one end these 2 Lectures were establish'd it seems by Sr: Henry Savill who has left the Professors also in Comon a Library whch they have there, and in it several fair Manuscripts copy'd from the Vatican by Savill himself and some more ancient particularly a very fair Euclid, He left them also a few Astronomical Instruments which are now decay'd and none of them to be us'd except a pretty good Quadrant. The Divinity school is only different from the rest being much more spacious, Lofty and handsome. Over the Schools, around three sides of the Square they make, you have a large Gallery with Bare Walls adorn'd however with several Pictures of the Founders of the several Colleges on one Side[199] and on the other with the Pictures of several very learned men & for the most part I believe Originals,[200] Among many others I remember I saw there Christopher Columbus, Martin Luther,[201] Dr: Bainbridge a famous Astronomer of Oxford[202] **[126]** Galileo,[203] F. Paul the Author of History of the Council of Trent,[204] Sr: Thomas Bodley, who gave their Library,[205] Chaucer the Poet,[206] a Map of the World in Arabick, Lord Leicester,[207] Selden,[208] Sr. Kenelm Digby,[209] Isaack & Merick Causabon,[210] Grotius,[211] ABps Usher,[212] Hevelius,[213] Dr. Wallis,[214] Sr. Hen: Savill,[215] the

199. These imaginary portraits of Founders dating from 1670 were probably by Willem Sonmans and were most likely commissioned by the colleges.
200. Kenneth Garlick, *Catalogue of Portraits in the Bodleian Library* (Oxford, 2004).
201. Martin Luther (1483–1546), presented to the university between 1704 and 1707; ibid., p. 209.
202. John Bainbridge (1582–1643), physician and astronomer, first Savilian Professor of Astronomy at Oxford University; ibid., p. 12.
203. Galileo (1564–1642), scientist and astronomer, portrait given to the university in 1661; ibid., p. 450.
204. Paola Sarpi (1552–1623), Venitian theologian, historian and 'eviscerator' of the Council of Trent; ibid., p. 272.
205. Sir Thomas Bodley (1545–1613) founded and endowed the Bodleian Library. Two portraits in collection; ibid., pp. 33–4.
206. Geoffrey Chaucer (*c.* 1340–1400), portrait listed by Thomas Hearne in 1705; ibid., p. 65.
207. Robert Dudley, Earl of Leicester (*c.* 1532–88), listed as in the Gallery before 1708; ibid., p. 112.
208. John Selden (1584–1654), scholar and writer. This picture, attributed to Sir Peter Lely, was hanging in the Gallery according to Thomas Hearne on 17 May 1708; ibid., p. 278.
209. Kenelm Digby (1603–65), Catholic virtuoso and a natural philosopher. Four portraits exist in the Bodleian collections; ibid., pp. 100–4.
210. Isaac (1559–1614) and Meric Casaubon (1599–1671), classical scholars. Both portraits were given to the university in 1674; ibid., pp. 53–4.
211. Hugo Grotius (1583–1645), scholar and jurist, portrait given to the university in 1674; ibid., p. 158.
212. James Ussher (1581–1656), Archbishop of Armagh, portrait bought in 1645; ibid., p. 307.
213. Johannes Hewelki (Hevelius) (1611–87), astronomer, portrait given to the university in 1679 by the sitter; ibid., p. 177.
214. John Wallis (1616–1703), mathematician, portrait by Godfrey Kneller; ibid., p. 353.
215. Sir Henry Savile (1549–1622), benefactor of the Bodleian collections, portrait given to the university in 1622; ibid., p. 273.

ABps Sheldon[216] and Laud,[217] in the same Gallery you see also a very large Cabinet of Medalls, but I believe not the most valueable in the World given by Abp. Laud with their Catalogues, you see also the Sword sent to Hen: 8th by the Pope along with the Bull I told you I saw in the Cotton Library for the title of Defender of the Faith, and in a Closet there you see two Pictures of Joseph Scaliger[218] and old Parr[219] who liv'd to above 160, An Oak chair made of the Wood of Sr: Francis Drakes Ship which Sail'd round the World, The famous Guy Faux's Lantern who was to have Executed the Gunpowder Plott[220] and a large Collection of Prints gather'd by Mr. Evelyn some of which are very valueable and now I think I have run thro' every thing I remember remarkeable in the Schools and the Library, Near to these you have the Museum – Ashmoleanum, which is worth visiting as well as the rest of it consists of one noble large room, round the Walls of which the several Curiositys are expos'd to view, they are indeed in great numbers but however I think they do by no means come up to Dr: Sloan's Musaeum in any respect much less in [**127**] their Rarity, Value or Elegant manner of Preservation, there is however a very exact Catalogue made of them which is annually compar'd with the Musaeum by 6 Persons appointed for that purpose among the heads of the Colleges, among the great variety of the things they shew you I remember some more particularly than the rest as for instance they have there a very large eagle headed Tortoise, they have a Picture of Sr: Elias Ashmole who was Founder of the Musaeum, a Cabinet of several little Artificial Baubles among which are several carv'd Cherry Stones, one I think with above 300 heads on it but very indistinct as you may imagine. A Little Pistoll not above an Inch long said to be Q: Elizabeth's Pocket Pistoll with which she amus'd her self in Shooting Flyes, They shew you in the same Drawers a small Horn exactly like a Ram's in all Respects but not above an Inch or two long which grew as they tell you out of the Forehead of a Woman in Saughall in Cheshire anno 1668 this is a great Curiosity but I am afraid with too good reason mix'd among the Artificialls, you have several other Cabinets of the Materia Medica, Shells and Specimens of Mineralls, a pretty perfect Mummy, an old Iron Scepter said to have belong'd to some of Our Saxon Kings, it is a very rude one & much of the taste of those times and seems to give rather an Opinion of Power than Politeness to the [**128**] hand that

216. Gilbert Sheldon (1598–1677), Archbishop of Canterbury; ibid., p. 282.
217. William Laud (1573–1645), Archbishop of Canterbury, portrait after Anthony Van Dyck and given to the university in 1674; ibid., p. 203.
218. Joseph Justus Scaliger (1540–1609), classicist. Both portraits are listed in 1679 in A. Wood, *The History and Antiquities of the University of Oxford* (Oxford, 1792).
219. Thomas Parr (?1483–1635), sesquicentenarian. Portrait made over to the university from the Tradescant collections by Sir Elias Ashmole in 1683 but since returned.
220. Given to Oxford University by Robert Heywood of Brasenose College in 1641, now in the Ashmolean Museum.

held it, a Statue of Marble Roman thought to represent Astrea tho indeed her Sword & Scales are so defac'd one can scarce be sure, a very large Roman Family Urn A great Congeries of Nails of Iron bigger than ones 2 hands melted into one Lump by Lightning yet so as plainly to appear what they were formerly, it seem'd to have been equally melted and consolidated within and without so that tho' they all Stuck in one Mass together one might I think blow thro' them in several places, besides these they have a Collection of about 4000 Medalls in this Musaeum but I believe not very valueable. I must not leave the room however without remembr'ing two very large Outlandish Beasts I saw there entire dead and their Skinns very well preserv'd by being stuff'd, one was a white Russia Rain=Deer, the other an Ethiopian Ass, I don't remember any where before to have seen whole Animalls thus preserv'd and so like nature too not even in Dr: Sloan's Museum and really I am much surpriz'd at it, methinks it might be a very easy thing by the Masters of Ships to procure the Skin and Limbs of most of the Quadrupeds and animals of the World dry'd in the Countrys where they are found and so brought over which would take but very little room, to be stuff'd here, I can't but think a Private Purse might very easily compass this much less a Prince's, and indeed I must own this would [**129**] by so much more noble and instructive a Musaeum than those that only preserve the lesser Animalls that I wonder some Prince has not attempted it. We see Gallerys adorn'd with the larger Productions of Art in Statues and Painting but nature in her more beautiful variety has no place. For my part I should have a greater Idea I am sure of a Prince to whom the World pay'd Tribute than of one confin'd to the narrow bounds of Greece and Rome, and I have no notion, why a Lyon from Africk, an Antilope or Hippopotamus, a mouse=Deer from America, a Northern Bear or an Eastern Elephant should not make as noble an Entertainment to surround a Sallone, as Pompey or Julius Caesar, At least I am sure for Variety this hint might be improved to make a most Glorious Gallery where there were others of Comon Sorts. On the Stairs of this Musaeum you see the Picture of Inigo Jones the Famous Architect and of the Tradescants a Family much addicted to these Studys of preserving nature in Museums one of which they made and you have a Catalogue of it printed. In this same building you have a natural Philosophy School and below a handsome Chymical Laboratory but not so good as Ours in Dublin College, and in a little room above you have a small Library where are some M.S.S. but I think none very valueable. They shew you there a Picture of Mr: Dee, who writt the fam'd book of Spiritts, another of [**130**] old Par and a Prospect of Mount Jenate a Burning Mountain I think in Africa. The Head Keeper of this Musaeum is one Mr. Parry a Welsh Man, who in Mr: Lhuyd's time was under him in that Office, this Gentleman is the same that accompany'd Lhuyd in his 5 years progress thro' Ireland, Scotland &.c. He told me that in Ireland they gather'd at least 1700 specimens of Mineralls which besides a great many plants Mr. Lhuyd kept in Cabinets which he shew'd me in a

Room there seal'd up some dispute having arisen between the University & one Mr. Price a Parliament Man of Wales who is now in the Temple a Lawyer in relation to the property of them, it seems this Price was Mr. Lhuyd's Executor and by virtue of that Will has already remov'd several old MSS from Oxford which Lhuyd got in his Travells and does now design also to remove all these Irish Mineralls and Plants to dispose of them for money which the University is not contented he should do Mr. Parry was so kind as to promise me however if possibly he could to procure me Specimens of each with marks severally where found at least he assur'd me I might depend on his sending me a Catalogue of them every one, which will contain many more than these mentioned by Mr. Lhuyd in his Lythophilacia Britannica, he gave me also some hopes of sending me an Exact Transcript of the Regular Journal he himself took in Ireland but assur'd me that Mr. **[131]** Lhuyd never made such a one at all and noted down now and then what he met remarkeable and that Mr. Lhuyd's Papers of that sort were so confus'd that he has not been able ever to make any thing of them since his death and therefore return'd them to the aforesaid Mr. Price, I am the more particular in this Affair because you know the promise Mr. Lhuyd in his life time made me of giveing me Copys of all his Notes in relation to Ireland which I should certainly have had but for his sudden death soon after; And here I cannot but observe to you one Custom that gave me great Offence at Oxford and this was the Schollars themselves shewing all their publick places for money and I am assur'd they have nothing else for their salary which is extreamly mean and low: I remember even Hearne took a piece from us at the Bodleian and here at the Museum we found the same Usage and indeed in every Library or almost Chappell we saw in the whole University. But now whether with greater Esteem or Regrett shall we approach the Muses awful Throne the Glorious Theater of Oxford, I had almost said Contempt as well as Regrett and think I may be excus'd when I tell you, that in the Niches round the Theater Court there are expos'd to the Air and weather the Arundelian collection & several other Marbles I believe about 100 in number & counted without doubt the finest Collection of Inscriptions, Relievos & Urns in England possibly in the World when they were **[132]** compleat there to be defac'd & to be destroy'd if it be not already done, that a University of their knowledge could have so little Taste or Thought itself is incredible for my part that have seen nothing, I griev'd at the Absurdity and could not bear to pass thro the Court but with reflection that they are at least preserv'd in Figures by Mr: Prideaux Book & are intended Suddenly but I fear for many of the Inscriptions too late to be remov'd within doors into the Room under the Museum. The Theater must compensate for the Transgressions of its Outside and Really I esteem this a very noble and proper room for its use it was the Gift of Abp Sheldon and the first reputed performance of Sr: Christopher Wren and it is very truly represented by

Loggan in his Mapps, it is built entirely of Hewn Stone in the figure of a long Semicircle 80$^{ft.}$ one way, 70 the other: round about the round end you have several Tires of Benches like a Roman Amphitheatre with proper places for Disputants, Orators, Proctors or others, that have to do in any of the Public Performances of the University and in the middle of the 4th or 5th Bench a Seat for the Vice=chancellor directly fronting the great door which is plac'd in the Middle of the flat side and has over it a Gallery for Musick, and at each side places for Ladys and Spectators besides the great open Area of the Floor, which will contain a great many [**133**] more. The top is painted to represent a Curtain gather'd quite round about to the Corniche exactly in the manner of a Roman Curtain they had to stretch across their Theatres to preserve them from the Sun, and in the open View you see the Muses descending into the noble room, and here indeed they dwell, here are all publickd Acts, Comitia, & Entertainments of the University celebrated, and I believe within these Walls as many illustrious performances & as much of the Pomp of Learning has pass'd as in any place whatever, in this respect I prafer it even to that which it emulates a Roman Amphitheater, and tho it falls infinitely short of their Stupendous Magnificence and Grandeur Yet when I reflect that this is a Scene of the Muses and that Apollo here praesides, while Mars & Gladiators fill'd those horrid Areas, I make no Scruple to look with greater pleasure & respect on the Amphitheater of Learning and Oxford. Below this noble room is the fam'd printing house of this University that place where so much Learning has had its rise and where no book has issued forth but in the most decent Editions and beautiful Characters, the Productions of this place are the most Cautious beautys, they never appear en deshabillee and seldom trust the greatest natural Charms without the Additional Set-off of the finest dresses in the World. They are now employ'd in raising a noble [**134**] Pile of hewn Stone near the Theater for a new printing house which will be yet vastly finer and more convenient. There are no other Buildings in Oxford that I esteem remarkeable except it be the University Church which is a great old Gothick building where all the Colleges meet at one Sermon every Sunday and another new Church of Hewn=Stone which is plain and handsome enough & call'd I think All–hallows; And thus when I have led you thro almost every Hall and Chappel, thro many of the Librarys particularly the Bodleian, when we have admir'd the Works and productions of nature in the Anatomy School and the Museums and those of Learning and Art in the Theater, when I tell you that I saw all these and consulted D$^{rs:}$ Plott & Wood about this University, when I found out two M.S.S they speak of one in the Bodleian a very ancient Arabian Geographer call'd Shariff Adrisi who mentions Oxford, and another in Magdalen College which mentions this University settled by Greeks probably those brought over about the year 668 by Theodor Abp of Canterbury, and when I tell you that all this was done in less than two days I am sure you will think I have given you a pretty good Account of this learned and Ancient University of Oxford and did not misspend my time while I was there, in effect I must confess I was extreamly pleas'd with my Journey so far & could have [**135**] heartily wish'd I had order'd my Affairs so as to have stay'd there a month or 2 which a particular View not only of the Genius of the Scholars & Colleges but even of the Librarys and buildings might very well employ I am confident. Before I return'd from Oxford to London you may be sure one day was employ'd in visiteing Bleinheim, 2½ h brought us there thro very bad Roads & worse weather so that we saw this vast pile and its Gardens to great disadvantage and could not help reflecting on the unhappy Season we saw it in. I send you enclos'd a very exact Plan of Blenheim and another of the Duke of Shrewsburys which is some 10 or 12 miles further and therefore we did not see it, this will explain to you the Situation of the Pallace and its Apartments better than I can possibly as I believe the Views, I send you of it from different sides will do for its different Fronts, I can only tell you 'tis entirely built of Portland Stone or Cornbury Stone; The grand Portico is in my opinion Clumsy and Crowded. The several Towers in the building extreamly Gothick and Superfluous, The great Face to the Avenue surprizeingly odd in my Opinion. The round Sweep of Pillars on either side the Portico not reaching up above half way of the whole height and the square building appearing backwards above it. The Statues too that are plac'd over these Pillars are very clumsy & ugly [**136**] and yet upon the whole I must confess this front at a Distance where its minuter Faults and virtues are not discernible does indeed appear noble and magnificent enough, The grand Avenue has a fall in it about 500 yards from the Steps of the Door of I believe about 20 or 30 Feet to the Terrain of the Bridge whereof I send you also a Figure enclos'd, This bridge indeed is extreamly large and noble and in my Opinion makes a very particular beauty arise to Blenheim from a Situation which otherwise had been very unhappy for where this Bridge stands there is a vast deep Vale, thro' which runs a little Stream directly across the Avenue & without this Bridge had entirely cut it off there Whereas now it rises gradually again on the other side and is continued with several Rows of Trees on a Plain to a very great distance I believe at least 2 or 3 Miles, The Arch I believe is at least 40 feet high and has in its Remparts and Buttresses all the Conveniencys of Water Engines to supply the house, bathing rooms and the like and I am assur'd will cost at least 1ˣ2000,[221] When you enter into the house I cannot but think the Hall appears extreamly handsome and magnificent 'tis of the Dimensions you see and I think 66 feet high, The Sallone is less & straitforward so as that these two Rooms make the whole thickness of the house in the Center,

221. This figure would appear to be £12,000 based on the 1714 estimate to complete the bridge of £6,994 4*s*. 2*d*. (BL, Add. MS 61354, f. 29). Molyneux may well have seen similar water supply systems at Kensington Palace.

for from the Sallone you step [137] directly into the Garden; on either hand of the Sallone you have two large state apartments of 4 rooms each but somewhat small so as that these nine you see make the Front backwards to the garden, the other two Sides are employ'd one in two private Apartments the other in a gallery from end to end I think 190 Ft: long, in this room you have several breaks and at about 30 or 35 ft: from each and two vast high Arches nearly as wide and high as the Walls and Cieling of the room so that in one view you have five different Breadths & heights at once and whether this will be agreeable when furnish'd or no I must leave the Connoisseurs to determine. Above these stately Apartments you have a great number of very convenient little Lodgeing rooms and some of them in little Appartments handsome enough and under all below you have vast arch'd Vaults and Cellars, places for Baths, Servants, Stewards & Chaplains very conveniently dispos'd; I must not forget to tell you that in one of these Halls below Stairs they shew'd Us a pair of very fine Tables of Course Agate and two Busts of the same, a fountain consisting of several Figures very well cut by the Sieur Bernini being a Modell in small of a Fountain in the Foro Aganali at Rome; it was made by him for the Marquis dell Carpi a Spanish Ambassador to the Pope and taken at Alicant but the most curious thing I observ'd there was a very fine antique Basso Relievo [138] of a great many Figures about 2ft: high & 5 long being the Face of some Cistern as appears by the Figures of two Lyons heads that are there represented with passage thro to lett the Water run out, this I think they told us was also taken at Alicant, and given to the Duke by General Gorges. On the East and West sides of this Pallace you have as well as to the back two Parteres one of which on the Side of the Gallery will be extreamly beautiful having on the Side of the house a Piazza under the Gallery of the whole length, and on the other a deep Valley, which after having Cross'd the Avenue turns & gives an Oportunity here of haveing a fine noble Canal in the bottom and very beautiful Slopes from the garden to it and the very agreeable Prospect of a fine riseing wooded hill on the other side which will add very much to the beauty of Blenheim Gardens & compensate for the plainness of the Gardens backwards which lye on a levell and have nothing in my opinion extraordinary in them at all, I can't however but observe to you that there is from the Garden door a view of 3 or 4 different churches at the ends of the walks which is pretty remarkeable because that these very views gave formerly before the Duke of Marlborough was born the name of Churchill to this very place as I myself have seen in several old maps. At every corner of this Garden you have Sort of Bastions and clear round a Course gravel [139] walk for a Chaize, From whence you have a good view of the Park, which is indeed very fine and as we were told so well stock'd with Deer that they seldom reckon less than 5000 brace at a time there. From these Gardens I must confess the Back front of Blenheim looks very well and if it were not for the Ugly Towers you see it loaded with I should have been much pleas'd with it. I shall not neglect to tell you that the whole house but the Hall and Gallery is already glaz'd and that entirely with Coach glass wch has been a vast expence in so large a house and no small Article in the sum of 3 ^{ld} 00000[222] which they say has already been lay'd out on this building tho' it be by no means finish'd not even the Shell and indeed I fear is like to continue so and to remain as little an Honour to the Nation as to the Builder that undertook it and I must confess I could not but agree with what a disaffected gentleman of our company say'd pleasantly enough of Blenheim tho I did by no means give into his insinuated Application that for his part take it all together, it had the very Air of a heavy Weight & Oppression on the Earth that bore it. If Woodstock be now the Seat of Valour & Beauty it has not long since known its Heroes & its fair ones. The Park was made a Mannor Royal by King Elfred who liv'd there and in this very place translated Boetius de Consolatione Philosophia; Henry the 1$^{st:}$ afterwards [**140**] built up the Park walls & had it for wild beasts Lyons and Tygers and the like as it is sayd in Story and perhaps divided it into the many little Paddocks it now contains. But Henry the 2nd is related to have put it to softer uses and that that in his praedecessor's time was a scene of hunting & Cruelty he made a Scene of Love and Happyness. The wildnesses and retreats that once had sheltered Savages were then but Labyrinths to his fair Mistress Rosamond, but where will not Jealousy find way more Savage and cruel than the Beasts before the Injur'd Queen enters the deep recesses and in the Rage of Jealousy destroys poor Rosamond; The tragick spot is still preserv'd and shewn to every body and near it a fine square pond of about 12 or 14 feet and 6 or 8 deep that bears the fair one's name and was her bathing place. How fair soever Rosamond was I am sure she was not more beautiful than this pond in the perfect clearness of its water and its agreeable Situation, it seems to have been formerly cover'd with a house, the Walls of which do partly remain and as we were assur'd lately several remains in Frescoe of ancient paintings now entirely defac'd; in the park you have an old lodge remaining which was part of an old Castle as 'tis thought of Henry's and that the Muses may not be wanting to adorn this Place, you are to know that its Heroes and Beautys do not more grace to it in their way than its Poetical [**141**] Inhabitants, for at the Park gate you see the old dwelling of Our Famous Chaucer who liv'd while he compos'd many of his Poems there. I do not doubt but you have seen a Figure of the famous Woodstock Pavement, a Roman Floor probably lay'd about Valentinians time of which Mr Hearn has given the

222. This figure of £300,000 appears only a slight overestimate on Molyneux's part. By the summer of 1712 all building had ceased with the costs incurred amounting to £220,000 with £45,000 owed to trademen and £13,000 paid to clear Woodstock Park. For more detail refer to Alan Crossley and C. R. Elrington (eds.), 'Blenheim: Blenheim Palace', *A History of the County of Oxford: Volume 12: Wootton Hundred (South) including Woodstock* (London, 1990), pp. 448–60.

World a Discourse in the 8th Volume of Lelands Itinerary, this is not above a Mile or 2 from Blenheim, and we therefore went to see it but unfortunately could not it being entirely cover'd over very deep with Earth to preserve it from the Winter weather so that without seeing it we were oblige'd to return to Oxford and I think met little or nothing remarkeable in the way except that I remember I saw a Sort of Plow lying on the Road the Beam of which was supported by an Axel tree between two Whells and of this I believe Dr Plott speaks of in his natural History as I am sure he does of Fairy Circles which I saw in several Fields that day exactly as he describes them. Monday the 23rd in the Evening we left Oxford and in about 2H., thro' bad Roads and a pretty woody Country we arriv'd at Abington which is a tolerable good large town and is very full of trade and Riches, They have here a handsome Town Hall and 2 or 3 very good little Hospitalls, the Revenues of one of which I was assur'd by one of the Governours was near 500ld per ann.

From hence the next day Tuesday 24 we went to [**142**] Dorchester and so to Henley, and, [gap]. On Wednesday the 25 along the Thames thro' a most lovely fine and well improv'd Country in Barkshire by Wallgrave [Walgrave], Twyvert [Twyford] and Hurst pretty Villages and by the Seats of Sr: Wm: Compton and one Mr Knight in about 4H to Maidenhead From whence on Thursday ye 26 we had a very easy but a very dirty Journ'y in 4½ h leaving Cliffden [Cliveden] a fine situated Seat of Lord Orkney on Our left and passing thro' Buckinghamshire, Stowe [Slough], Colebrook [Colnbrook], Hounslow & Brentford – several Villages to the August City of London And thus compleated Our Visit to Windsor; Oxford & Blenheim exactly in a Week.

I am Yrs Sincerely,

A Postscript in this letter: I think may be very well employed in letting you know I have been Since I came to Town to see the very valueable Collection of Antique Arundelian Gems which are now in the hands of Sr John Germain, I send you herein also enclos'd a Catalogue of them and can only tell you that I compar'd it with the Gems and find it very exact and am fully satisfied of what several have assured me to be true that there is [**143**] not such a Collection in the World so good work so valueable so many and so well preserv'd. The Pope and some others are treating about buying them but I am in hopes my Lord Treasurer who has been apply'd to on that head will not suffer them to go out of England when so small a sume as 6000ld is their purchase. I must not forget to tell you that while we were looking on these Gems by the Assistance of our good Friend my Lord Pembroke he happen'd to speak of Mr Pitts his great Diamond of which he has seen the Model, he gave us this Account of it that it was cut into a Brilliant of a long square shape and weigh'd 138½ Caratts of a perfect clear Water he gave me no hopes even of seeing the Modell which they are extreamly cautious in produceing on any account, to confess a truth I could not but reflect with some Surprize on the vast worth of the former artificial Curiositys & on the much more imense value that vanity has sett on this natural one which according to the Comon manner of valueing Jewells will amount to about 2 millions. I have been also since the Conclusion of this letter very well diverted in seeing two Curiositys which I hardly thought worth visiting till I saw them. The one of these was Bridewell or the house of Correction belonging to Westminster, the house itself is however scarce worth mentioning and I had not troubled you with speaking of it were it not to tell you that it stands encircled with several Almshouses [**144**] and Charity Schools in which above 300 people are maintain'd by private gifts of Charity from several people in all Circumstances of life and many more educated. It seems the Land there belongs to the Dean and Chapter of Westminster who giving it out gratis to these charitable Foundations have occasion'd there being crowded here altogether, and indeed it was with some pleasure we reflected that the same place contain'd the greatest Helps to Virtue and good Education and the sharpest Punishments of Vice and Wickedness. The other Curiosity that I have seen is the Museum of Mr Petiver an Apothecary in Aldersgate Street which is really in its way very well worth seeing it consists principally of several Volumes of dry'd Plants amounting in number to above 20000 from all Countrys of the World as also of most of the known Insects. I do particularly esteem the method of his Collections which is in all its parts digested according to the different parts of the world so that you have the Plants and Insects of Europe by themselves and so of the rest, The Vegetable & Animal World indeed of Brittain has a particular place and very properly, for I think all Collectors of this kind should have a particular regard to their own native Country. The principal Curiositys I remember there were some Cantharides found in England, which it is hop'd may in time propagate so as to be useful. The true ancient [**145**] Ancardium Orientale from Malabar a Plant much us'd by the Ancients as a very excellent Medicine and frequently mention'd in their Writeings tho' now scarce known except from the West Indies where they have not the true sort, A Dens Moharis of the Animal call'd Chamans from the West Indies petryfy'd and which is much more surprizeing a little Snake rowl'd up and petryfy'd or rather incrustated with Stone taken in the Stomach of a Calf in Norway from whence also he shew'd us several other instances of like petrifications taken in the bodys of Animalls in such Shapes & Consistence as do plainly evince notwithstanding Dr: Woodwards Hypotheses that Petrification can be made in very different manners from what he supposes, He shew'd us however above 40 Sorts of East and West India Shells found in one bed all heap'd together at Limmington in Glocestershire, This Gentleman is a very laborious Collector of Nature and I believe if his Education had been suitable to his Genius & Inclination he might have been very useful to the Enquireing world, He observ'd to us that he finds several Chinese plants very distinctly represented in the Comon India painted Paper & assur'd us that in their designing where there was no perspective to be represented they have not been so rude as is generally Esteem'd, The Analogy also which he observes as he assur'd us in the Virtues of those plants that agree in Shape, & [**146**] Colour is also very remarkeable. I saw also the other day two very fine pieces of Mechanism in the hands of Mr. Roberts one was a clock made at Nuremberg, the Pendulum of which was regulated by a little Ball, rowling down several declining Plains, and when arriveing at the bottom return'd again to the Top by one half Gyration of the Pendulum or Regulator and this measures equal times he says most exquisitely: The other was a square Chest or Strong box the Key of which was a Skrue & being Skrued in at length press'd on some spring which made all the bolts fly back which fasten'd the Box above 40 in number by all shooting to the Center out of the sides of the Box into as many Staples in the Lid. This was made in London and an invention of Prince Ruperts I think.

LETTER 6

(complete: ff. **146–54**), 15 April 1713

SCA D/M 1/3

This, the sixth letter in the copy-book, is complete and looks back over Molyneux's time in the capital which he says has amounted to 'above 5 months' (f. **154**). By the time the letter is completed on 15 April 1713, Molyneux's departure for Holland is imminent. On the same day he leaves London for Cambridge, en route to the busy port of Harwich in Essex, from where, soon after 20 April, he sails to Ostend to join the Duke and Duchess of Marlborough.

Although the content of this letter, on the whole, has a more social bias, for the benefit of his reader he encloses a comprehensive list of the 'Curiositys' that he has visited since arriving in London the previous October. Unfortunately, that list has not survived, but in the letter he also commends the hospitality of men of 'distinguish'd Character' that he has '[...] had the Honour since I have been here to be sometimes receiv'd by' (f. **148**), compiling an orderly list arranged according to rank and hierarchy. Beginning with peers of the realm and then moving on to theologians, writers and scientists, the list reads 'the Duke of Marlborough, My L$^{d.}$ Pembroke, My L$^{d.}$ Hallifax, the Duke of Devonshire, my L$^{d.}$ Sunderland, the Archb^{p} of Canterbury, the Bishops of Ely, Bangor & Litchfield, M$^{r:}$ Addison, M$^{r:}$ Steel, M$^{r:}$ Congreve, D$^{r:}$ Swift, M$^{r:}$ Phillips, S$^{r:}$ Isaac Newton, D$^{r:}$ Sloan, M$^{r:}$ Halley' (f. **148**). This record of names is useful in filling in some of the gaps in the account of Molyneux's engagements that have been lost from the manuscript due to the intermittent missing folios.

First and foremost on the list of names is the Duke of Marlborough, who was about to be (if not already) Molyneux's patron. Where and when they first met is not confirmed, nor does Molyneux expand on where he encountered Lord Halifax, despite giving an account of Halifax House (during his visit to the Palace of Westminster, ff. **42–3**) and of Bushy Park where Halifax was Ranger (f. **86**). Thomas Herbert, Earl of Pembroke, is first mentioned by Molyneux when he visits his collections housed at St James's Square 'which', Molyneux writes, 'I have seen with as great Pleasure as my Lord shews them and this I assure you is so good natu'rd & loves so much to instruct a Stranger he does with the utmost willingness and Satisfaction' (ff. **79–80**). Molyneux appears again with 'our good Friend my Lord Pembroke' (f. **143**) whilst viewing the gem collection of Sir John

Germain, one of the leading virtuosos of the day and a close neighbour of the earl's. Another important collection he mentions is that of the prominent Whig and supporter of the Hanoverian succession, William Cavendish, the 2nd Duke of Devonshire (1672–1729), whom Molyneux may have met at Devonshire House where the duke kept his collection of coins, engravings, gems and paintings (some by Titian, Rembrandt and Van Dyck). Charles Spencer, the 3rd Earl of Sunderland (1675–1722), was, like Devonshire, a political supporter of the Duke of Marlborough and staunch Whig who challenged the Lord Treasurer, Robert Harley, for the position of the greatest bibliophile of the day. His library 'running from the House into the Garden' at his home, Sunderland House (Map I㉒) (since demolished) in Piccadilly was described in 1714 by John Macky as

> [...] the Finest in *Europe*, both for the Disposition of the Apartments , as of the Books. The Rooms, divided into five apartments, are full 150 Feet long with two Stories of Windows, and a Gallery runs around the Whole of the Second Story, for taking down the Books. No Nobleman in any Nation hath taken greater care in making his Collection complete; nor does he spare any cost for the most valuable and rare Books. Besides, no Bookseller in *Europe* hath so many Editions of the same Book as he; for he hath all, especially the Classicks.[223]

Following on after the aristocratic collectors, Molyneux lists the Archbishop of Canterbury (Thomas Tenison, 1636–1715) and the bishops of Ely (John Moore, (1646–1714) (Map II㉗), Bangor (John Evans, 1651–1724) and Lichfield (John Hough, 1651–1743). It is tempting to suggest that Molyneux's motive for associating with these senior churchmen is to engage in learned debate about the corrosive effect of natural philosophy on religion; however, it would appear more plausible that theology *per se* is not the attraction, rather the collections owned by these great men. Archbishop Tenison was a strong supporter of the Hanoverian succession and became one of three officers of state who appointed a Regent until the arrival of King George I, whom he crowned on 20 October 1714. A notable bibliophile of the day, Tenison built London's earliest public library and endowed it with a fine collection of books and manuscripts.[224] It is likely that Molyneux visited the archbishop at Lambeth Palace (Map VI㉓). John Moore, Bishop of Ely between 1707 and 1714, was also a great collector of books and manuscripts which were said to be 'universally and most justly reputed the best furnish'd of any (within the Queen's Dominions) that this

223. Macky, *A Journey through England*, p. 192. The collections were transferred to Blenheim Palace by Sunderland's son who became the 3rd Duke of Marlborough.

224. Peter Hoare, 'Archbishop Tenison's Library at St Martin-in-the-Fields: Its Building and its History', *London Topographical Record*, XXIX (2002), pp. 127–50. Lambeth Palace Library was, in fact, older but was not generally available for consultation.

Age has seen in the hands of any private Clergyman'.[225] His thirty thousand volumes were housed at his episcopal house in Ely Place, Holborn, 'in eight chambers [...] that almost surround the quadrangle'.[226] On Moore's death in 1714 King George I purchased the collection and donated it to Cambridge University where it became known as the King's or Royal Library. It is less obvious why Molyneux should have come into close contact with John Evans, Bishop of Bangor between 1701 and 1715, or John Hough, Bishop of Lichfield and Coventry between 1699 and 1717. Evans was a staunch Whig and highly-principled man who became Bishop of Meath in 1714, while Hough was a less committed Whig but, having been based in Ireland during part of the 1680s, it is conceivable that he knew the Molyneux family.

Next, Molyneux names members of a select literary circle that included Joseph Addison and Richard Steele (*c.* 1672–1729). Of Addison, Molyneux writes of 'the Friendship I have for [the] Author' (f. **152**) which suggests that their acquaintance may date from April 1709, when Addison was based in Dublin as secretary to the Lord Lieutenant of Ireland, Lord Wharton. Addison and Steele were active Whigs and prominent contributors to periodical literature, most notably *The Tatler*, which ran between 1709 and 1711, and *The Spectator*, which ran for 555 editions before closing on 6 December 1712 while Molyneux was in London. Neither publication is mentioned by Molyneux nor, with the exception of Alexander Pope's poem *Windsor Forest* (f. **111**), does he demonstrate any particular literary interests. He does, however, comment on one of the cultural highlights of his stay in London, when he attends the first night of Addison's play *The Tragedy of Cato* (f. **152**), a political tragedy inspired by the Roman hero, that was premiered at Drury Lane (Map II ㉖) on 14 April 1713 and ran for twenty-eight nights to great acclaim. Molyneux's enthusiasm for the play may well arise from some affinity with the Roman hero, of whom *The Spectator* wrote

> [...] that since the human Soul exerts it self with so great Activity, since it has such Remembrance of the Past, such a Concern for the Future, since it is enriched with so many Arts, Sciences and Discoveries, it is impossible but the Being which contains all these must be Immortal.[227]

Addison and Steele may well have provided introductions to the playwright and poet William Congreve (1670–1729) and Ambrose Philips (1674/5–1749), who wrote *Pastorals* in 1709, a work heartily applauded by *The Spectator* but heavily criticised by Alexander Pope and parodied by John Gay in *The Shepherd's Week* in 1714.[228]

225. W. Nicolson, *The English Historical Library* (London, 1714), p. xii.
226. Hunter, *Ralph Thoresby*, p. 116.
227. *The Spectator*, No. 537, Saturday 15 November 1712.
228. *The Spectator*, No. 523, Thursday 30 October 1712.

Last on the list are the great men of science who had most influence on the young Molyneux in his mission to become a natural philosopher. Sir Isaac Newton (1642–1727), the President of the Royal Society, and the astronomer Edmond Halley (1656–1742) had, by the time of Molyneux's visit, controversially published John Flamsteed's research in *Historia coelestis* and were working together on a second edition of *Principia* which reassessed Newton's notions on lunar theory. The ruthless and determined Newton must have known of Molyneux (and presumably admired his work) before his arrival in London, as it was he who nominated the Irishman for a Fellowship of the Royal Society. Perhaps it was at the headquarters of the Royal Society, recently moved from Gresham College to Crane Court, off Fleet Street, that Molyneux met these two scientific giants. Indeed, a description of this meeting may have formed part of the missing folios. The other scientist Molyneux cites is Dr Hans Sloane (1660–1753) (Map II ㉔), whose 'curiosities' were described by John Evelyn in 1691 as 'an universal Collection of the natural productions of Jamaica consisting of Plants, Corralls, Minerals, Earth, shells, animals, Insects &c: collected by him with greate Judgement, several folios of Dried plants & another of Grasses: &c'.[229] The museum was partially formed from a bequest by William Courten (1642–1702) (whose own museum was held at Middle Temple), but by the time of Molyneux's visit had been added to significantly. In 1698 the collection had moved to Bloomsbury Place where Ralph Thoresby wrote, '[Sloane] has a noble library, two large rooms, well stocked with valuable manuscripts and printed authors, an admirable collection of dried plants from Jamaica, the natural history of which place he has in hand [...] he has other curiosities without number, and above value'.[230] It is certain that Molyneux visits Sloane's museum as, when writing up his visit to the Ashmolean Museum in Oxford, he states, 'however I think they do by no means come up to Dr: Sloan's Musaeum in any respect much less in their Rarity, Value or Elegant manner of Preservation' (ff. **126–7**). For privileged visitors Sloane presided over the tour himself, which may suggest the meeting with Molyneux.[231] The collection later formed part of the foundation of the British Museum.

In addition to his fascination with the controversial High Church preacher Dr Henry Sacheverell, further curiosities he witnesses are another (unnamed) 'Dr of the church' (f. **148**) who possesses remarkable teaching powers, the Swiss memory man 'Mr Heideker' (f. **149**) and Alexander Selkirk, who was

229. de Beer, *John Evelyn*, p. 827.
230. Hunter, *Ralph Thoresby*, p. 299.
231. See Marjorie L. Caygill, 'From Private Collection to Public Museum: The Sloane Collection at Chelsea and the British Museum in Montague House', in R. G. W. Anderson, M. L. Caygill, A. G. MacGregor and L. Syson (eds.), *Enlightening the British: Knowledge, Discovery and the Museum in the Eighteenth-Century* (London, 2003), pp. 18–28.

soon after immortalised by Daniel Defoe as 'Robinson Crusoe'. Selkirk (1676–1721) created a great deal of interest in London after his ordeal at sea and was the subject of two contemporary books.[232]

Whilst writing about some of London's celebrated inhabitants and their trades, Molyneux makes a specific point of mentioning the bookseller and auctioneer Christopher Bateman, whose shop was situated at the sign of the Bible and Crown (Map III ㉕) on the corner of Ave Maria Lane and Paternoster Row. Jonathan Swift is known to have frequented Bateman's, writing,

> I was at Bateman's the booksellers, to see a fine old library he has bought; and my fingers itched, as yours would do in a china shop; but I resisted, and found everything too dear, and I have fooled away too much money that way already.[233]

Another contemporary commentator added,

> There are very few booksellers in England (if any) that understand books better than Mr. Bateman, nor does his diligence and industry come short of his knowledge. He is a man of great reputation and honesty, and is the son of that famous Bateman, who got an alderman's estate by bookselling.[234]

In his *magnum opus* on Archbishop William King's extensive library in Dublin, Robert Matteson writes that 'Dublin, London and Bath were the principal centres for King's book-buying and, as with any collector with a daytime job, King had his agents — John Baynard, Francis Annesley and Samuel Molyneux'.[235] Molyneux's book-dealing on behalf of the archbishop is mentioned in the correspondence between the two men and referred to in the Appendix.

This lively letter is quite unlike any of Molyneux's previous correspondence as here, for the first time, he momentarily departs from elitist London society in order to focus on the wider social and cultural life of the city. His view of London as a socially diverse metropolis and a multicultural

232. Captain Woodes Rogers, *A Cruising Voyage round the World. Begun in 1708, and Finished in 1711. Containing an account of Alexander Selkirk's living alone for four years and four months on an island, &c*, (London, 1712) and Captain Edward Cook, *A Voyage to the South Sea and around the World. Wherin an account is given of Mr Alexander Selkirk, his Manner of Living, and taming Wild Beasts, during the four Years and four Months that he lived upon the uninhabited island of Juan Fernandez*, 2 vols (London, 1712).
233. Henry Clinton Hutchins, 'Dean Swift's Library', *Review of English Studies*, 9, no. 36 (October 1933), pp. 488–94.
234. Henry R. Plomer, *A Dictionary of the Printers and Booksellers who were at work in England, Scotland and Ireland from 1668 to 1725* (Oxford, 1922), p. 24.
235. Robert S. Matteson (assisted by Gayle Barton), *A Large Private Park: The Collection of Archbishop William King 1650–1729*, 1 (Cambridge, 2003), p. xiii.

centre comes with the caveat that 'as I believe in most great Societys of Mankind you meet the most learned men and the greatest Rogues' (f. **147**). It is noticeable that Molyneux does not explore the sinister or roguish side of the capital beyond the coffee houses and the popular press, suggesting either that he never fully experiences London's menacing side or, more likely, that he guards his comments carefully to avoid worrying his guardian uncle.

It is not clear if Molyneux attends the opera whilst in London, though he does portray its decline in popularity during the winter of 1712/13. By 1710, the all-sung Italian opera was well established in the capital and the celebrated castrato, Nicolini, was one of its foremost practitioners. In 1709, *The Tatler* wrote of Signor Nicolini

> [he] sets off the character that he bears in an Opera by his Actions, as much as he does the Words of it by his Voice. Every Limb, and every Finger contributes to the Part he acts, insomuch that a deaf Man might go along with him in the Sense of it. There is scarce a beautiful Posture in an old Statue which he does not plant himself in, as the different Circumstances of the Story gives Occasion for it. He performs the most ordinary Action in a Manner suitable to the Greatness of his Character, and shews the Prince even in the giving of a Letter, or dispatching of a Message.[236]

The Spectator concurred and described him as 'the greatest Performer of Dramatick Musick that is now living, or that perhaps ever appeared on a stage '.[237]

Although Molyneux regularly describes his own encounters with polite society, he rarely refers to the wider social life of the city. This letter, however, refers to two contrasting aspects of London society: the coffee-house and the learned society. Molyneux blames coffee-houses and the press for fuelling much of the civil unrest in the city, and for many the two were perceived to be inextricably linked to the spread of malicious gossip and falsehoods. Tom Brown described the coffee-house in 1700 as a place where

> [...] knights-errant come to seat themselves at the same table without knowing one another [...] they have scarce looked about them, before a certain liquor as black as soot is handed to them, which being foppishly fumed into their noses eyes and ears, has the virtue to make them talk and prattle together of everything but what they should do.[238]

The Spectator provides a good description of the significant coffee-houses in London, at the time of Molyneux's visit, as having '[...] some particular

236. *The Tatler*, No. 115, Saturday, 31 December 1709, pp. 1–2.
237. *The Spectator*, No. 405, Saturday, 14 June 1712.
238. Tom Brown, *Amusements, Serious and Comical for the Meridian of London and Letters from the Dead to the Living* (London, 1700).

Statesman belonging to it, who is the Mouth of the Street where he lives, I always take care to place my self near him, in order to know his Judgement on the present Posture of Affairs'.[239] Judging by Molyneux's dismissive comments, it is unlikely that he patronised the 'Grecian' in Devereux Court, near St Paul's Cathedral (the favoured establishment for members of the Royal Society) or St James's coffee-house which was frequented by gentlemen from Ireland.

In direct contrast to the coffee-houses and dubious reporting of Fleet Street was the learned society, where intellectual conversation between men of 'good Sense and [...] distinguishe'd Learning' (f. **154**) would take place. Through his uncle's recommendation, Molyneux attends a weekly Monday night meeting of a society of 'Belles lettres' (f. **154**), of which he is immediately made a member. With a membership of 'principally [...] Foreign Clergymen', the meetings take place at the house of fellow Royal Society member Louis Frederick Bonet, the King of Prussia's Resident (or Ambassador) in London.

The letter ends somewhat abruptly with Molyneux's announcement that he is about to depart for Cambridge, a journey which the next letter makes clear began at 10 a.m. on 15 April 1713.

Text of Letter 6 (complete)

[**147**] Dr: Sir London April 15th 1712/3

Having now given you all the Observations I have made on what is to be seen in London as to its particular Curiositys I believe one closeing letter from this place may be very properly employ'd in letting you know the general Remarks I made in this City, to do this with any kind of Exactness may be perhaps be very tedious & difficult, I shall therefore venture no further than to tell you here and there some of the most observable things that occur at present to my Memory and here the most distinguishable is certainly the great Populousness and variety of the Inhabitants; you may I think in London visit Europe and meet natives of all the Countrys in the World, the Temperateness of its Climate, makes it as easy a Retirement from the Sands and Heats of Affrick as from the Ice and Desarts of the North. Graecians and Sweeds meet in its Streets and find their Countrymen already Settled there, nor are its Inhabitants in this respect less various than in many other extreams, it is here as I believe in most great Societys of Mankind you meet the most learned men and the greatest Rogues, The deepest Politicians and the most trifling Fops, the most beautiful insides in the most ordinary [**148**] Aspects and very much too often the Reverse. The great and Happy Liberty of England has made the Noblemen more than commonly civil & Familiar and I must say this

239. *The Spectator*, No. 403, 12 June 1712.

for the Politeness of London that I have always met since I have been here with greater Affability and Courteousness from People of the most distinguish'd Character than I could even hope for from those of my Sort, to give you the best Argument I can of their great Condescension I need only tell you, that I have had the Honour since I have been here to be sometimes receiv'd by the Duke of Marlborough, My Ld. Pembroke, My Ld. Hallifax, the Duke of Devonshire, my Ld. Sunderland, the Archbp of Canterbury, the Bishops of Ely, Bangor & Litchfield, Mr: Addison, Mr: Steel, Mr: Congreve, Dr: Swift, Mr: Phillips, Sr: Isaac Newton, Dr: Sloan, Mr: Halley, besides these and a great many others with whom I have had the Honour to be made acquainted, I must not neglect to name to you several other very remarkeable Men I have met in London, among these I am certainly to name the Fam'd Dr: Sacheverell whom I had the Honour to hear read Prayers, and must say that if his Doctrine were agreeable to his voice they would be full of the softest and most effectual Perswasion, another Dr: of the Church that I can't forget is one [gap] who employs himself in teaching to write and that so as in 19 hours time to perfect a Comon Scholar in any hand whatever, this Art is somewhat **[149]** incredible and I could scarce venture to mention it but that I assure you he produces several instances of his success in this attempt I hear his method is to perfect you in the exquisite true formation of one letter before he begins with another whereas other Masters do not do so I believe his method is only the most easy that has been thought on. We have here a Foreigner who is no less remarkeable than this Gentleman in having a prodigious and very odd contriv'd memory one Mr. Heideker a Swise, he will remember above 100 incoherent words, the Signs of a Street from End to End or Mens names or the like and this I have been assur'd by several ingenious Men who have seen him do it and have been told by him that it is a peculiar Art he has and no natural Gift and that the more incoherent the Words he finds it the easyer to remember them and could not with all his Skill get a Speech by heart. I remember also since I have been in London I once saw Mr: Alexander Selkirk who liv'd four years in a desert Island alone without houses or Provisions left him. I saw a Gentleman that had been 5 years a Prisoner in Algiers & another that liv'd 3 years in the deserts of Arabia amongst the wild robbers. That part of London that has been burnt is now rebuilt into very beautiful good Brick houses generally Tyl'd and of those the whole Citty within the Walls is generally compos'd and indeed besides the Burrough of Southwark **[150]** which is on the other side the Water and has several very good houses and two Extraordinary fine Hospitalls which deserve very well to be seen; there are dispers'd in Westminster and the Suburbs very many beautiful Squares & good houses particularly I am oblidged to name one in Lincolns Inn fields belonging to one Coll: Child which I was carry'd to see with great pleasure, the Elegance of its Furniture & the goodness of the Pictures do I think give it the Praference above all the private houses

I saw in London; The Halls of several of the Corporations particularly Merchant Taylors and Mercer's Hall are very handsome & I have seen them with great Satisfaction. The great Opulency and Trade of this Citty has unquestionably been the occasion of such large Publick buildings and indeed this appears in a 100 respects more, The largeness and Riches of their Shops, the particularity of their Trades I mean that the Artificers here find such great demands of their Manufactures & so sure a Vent that you shall have ten different Trades where in Ireland we have but one, you shall have a Glass Framer for Coaches that makes it his only trade and yet getts enough too, another for the Harness & so in all other Trades which I take to be a certain Sign of Riches, The very signs in London are remarkeably rich especially in the great Street and I am tempted to think the very signs here would near buy the Shops in Dublin. I am sure to throw in the Rails of their [**151**] Houses which with the Supports of the Signs are all Iron work and very good it would come very near the Purchase, when I mention the Opulency of Shops I must not forget my good Friend Mr: Bateman who I believe is the best Bookseller for choice Editions and old Librarys in the whole world and has accordingly vast Loads of Books in his Shop at all times. The Number of Coaches and Carts in London are also vastly surprizeing and indeed there is no time of Day you can pass the great Street without meeting crowds of People & Coaches & Carts to Stop you, the Carts are made with vast heavy large Wooden Wheels unshod on which the Body of the Cart is nearly pois'd as high as the Horses back, the great Streets have generally at each side two flagg'd Walks of 8 or 10ft wide separated by Posts from the Coach way; The Lights indeed of London both in the Citty and Suburbs are very few and not well dispos'd which is extreamly inconvenient and I don't remember any one Street, but the one I lodg'd in Suffolk Street so well enlightend'd as all Dublin is. The manners of liveing in London are very different; In Wapping suburb you have nothing but Seamen, in the citty Trade, Politicks & Riches, insomuch that I am assur'd by one that certainly knows London is ¾ths of England for Trade, In Westminster and St James's Suburb, Pleasures, Witt & Diversion, In the Citty your eating & Lodging and every thing is in a Track [**152**] of Thrift and saveing, at the other end of the Town you meet all things prepar'd for People of Extravagance and Luxury, of these Entertainments the Opera was so extraordinarily perform'd the last Winter, and while Signore Nicolini was in England that indeed this Winter they have been in a very bad Condition, 2 or 3 ordinary Pastoral and one or two Operas are all that have been perform'd and these by no means as they tell me in Perfection. However indeed the house and Scenes are very fine. The falling-off of the Town from the Taste of Operas has been occasion'd I believe also in a great measure by the Recommendation of the much more Rationall diversions of Plays given them on all Occasions by our present sett of

Witts and indeed it is with great Justice they stand up in defence of the Dramatick Theatre when they are able to divert the Town with such finish'd pieces of Poetry as the Tragedy of Cato which I last night saw perform'd to my great Satisfaction as well in Regard to the Friendship I have for its Author as to the good and virtuous Taste of the Town with the utmost Applauses possible, and I do not doubt but it will be able to Spirit up the World into a new Sense of Roman Magnanimity & Love for their Country. It is with great Regret I must reflect on this Occasion on the very noble and well tasted diversion I have left behind me at York Buildings, You know Our Friend there designs in a Theatrical Representation of the most beautiful & instructive Passages & Speeches of the Antients, visibly to [**153**] Cheat us into a new Scene of True Virtue and Greatness of Thought and has accordingly fitted up at vast expence a room for this purpose with places for about 200 persons. I don't know whether London has now so many of refin'd and dareing taste as to compleat such an Audience but I am confident if it has not it soon will and that no Project can be of such use in makeing such an Audience as this. It is with no less Regret I must reflect on the great height of partys now in this place, The happy Liberty of England by the vast Temptation of its many Offices of great value and honour is debauch'd continually into these Misfortunes and while places in great Brittain are estates & Foundations of Familys and Nobility those that wish for them will never cease to make use of the Passionate Love of Liberty to make odious one anothers party to the People by aggravateing & misrepresenting each others designs and Schemes of Administration, and thus we are like to continue for ever in flames. The Licence of the Press is a great Cause of these Mischiefs. The News writers and the Number of Coffee-houses, above 3000 in London, tho' not 50 years agoe a man was indicted for vending that unwholesome Liquor; these also I say I take to have a Share in these Affairs, but how it is possible to apply a Remedy without invadeing the People's Liberty is what I confess I am not able to contrive and can Rather [**154**] than hope for. I can remember nothing more in this Miscellaneous way than what I have told you of London except it be to let you have an Account of our Munday night conference as you desir'd; This Assembly then consists principally of Foreign Clergymen, there are some Laymen of which M[r:] Bonet the Prussian Resident is one, and at his Lodgeings we met, one Gentleman proposes a question to be treated the next week and accordingly opens the Conference by a discourse on that Subject, and after that every body speaks in their clear round their thoughts extempore. The Subjects are generally Moral of the Belles Lettres and indeed I must confess since they did me the Honour of makeing me one of their Members I have always been extreamly well pleas'd and instructed in their Members, for I assure you all my Brethren are men of Extraordinary good Sense and most of them of distinguishe'd Learning, and thus I finish my long troublesome Account of this great citty, I think I have been above 5 months here, and tho I am of Opinion one month might serve to see all its Curiositys, yet I assure you I have passed my time very well & have seen some of them with Pleasure much oftener, I send you enclos'd the List I went by in my Visitts of these Curiositys and have found it pretty I just step into the Coach for Cambridge & thence directly to Holland

& am Y^rs Sincerely

LETTER 7

(incomplete fragment: ff. **156–9**), 20 April 1713

SCA D/M 1/3

Molyneux's journey from London to Cambridge takes him through Bow and past Wanstead House (now demolished), the 'very fine seat' of Sir Richard Child, the 3rd Baronet of Wanstead.[240] After taking lunch at Epping, he and his travelling companions journey past the home of Lieutenant-General William North, Lord North and Grey (1678–1734), before spending the night at Bishop's Stortford. Making an early start on 16 April, they continue onwards past Audley End, the home of Henry Howard, the 6th Earl of Suffolk, who, from 1708, demolished much of the old-fashioned house. On arrival at Littlebury, he visits Littlebury House — the self-styled 'House of Wonders' — where he enjoys the curiosities of its builder, the engineer, artist and eccentric inventor Henry Winstanley (1644–1703). After Winstanley's death the house and collection had been kept intact and opened to the public by his widow, Elizabeth. Molyneux is drawn to Winstanley's ingenuity as an inventor and particularly enjoys the mechanical chairs. Also on display is a model of the Eddystone lighthouse. In 1698, after two years of building, the Eddystone lighthouse in the English Channel near Plymouth was completed. But in 1703, after some problems with the tower and defiant in his belief that the structure was sound, Winstanley lost his life when the lighthouse collapsed in one of the worst storms ever to hit the coast of Britain.

In July 1710, Zacharias von Uffenbach had followed the same route, writing that 'we set out at 8 a.m. from Littlebury, and reached Cambridge before 12, a distance of ten English miles'.[241] Cambridge, he added, 'is no better than a village [...] the inns, of which there are two, are very ill-appointed and expensive', claiming that in his view 'were it not for the many fine colleges here it would be one of the sorriest places in the world'.[242] Like von Uffenbach, Molyneux visits Trinity College first but, unlike the German, Molyneux dines in the Hall which von Uffenbach had described as 'very

240. Howard Colvin, *A Biographical Dictionary of British Architects 1600–1840* (Yale, 1995), p. 211. Wanstead House, the magnificent Palladian mansion, was designed by Colen Campbell.

241. David C. Douglas (ed.), 'Zacharias Conrad von Uffenbach on the Universities, 1710', *English Historical Documents 1660–1714* (London, 1953), p. 486.

242. Ibid., p. 489.

large, but ugly, smokey, and smelling so strong of bread and meat that it would be impossible for me to eat a morsel in it'.[243] Molyneux's privileges arise from his letter of introduction to Mordecai Cary (1687–1751) who entered Trinity in 1705, received his M.A. in 1712 and was ordained as a deacon in London in 1714. It is not clear why Molyneux's guide at Trinity is Cary, but it could be that the letter of introduction was intended for Cary's tutor, James Jurin (1684–1750), who was a follower of Newton and protégé of Richard Bentley, Master of Trinity between 1700 and 1742. Molyneux thinks the chapel of Trinity College 'very handsome' (f. **158**), referring admiringly to the recent works commissioned by Bentley. In the interests of advancing science within the College, Bentley added a new chemical laboratory for Giovanni Francesco Vigani, the first Professor of Chemistry at Cambridge University, and a new observatory which Molyneux notes as still being under construction above the King's or Great Gate. The new observatory was approved by the master and seniors on 5 February 1706, but progress was slow due to a lack of finance — it was finally finished in 1739, but was reported as being out of use for over fifty years in 1792 and demolished in 1797. It is not clear if Molyneux actually meets the 'Professor of Mathematicks and much esteem'd' (f. **158**) Roger Cotes (1682–1716), but they certainly had a mutual acquaintance in Sir Isaac Newton. Newton recommended Cotes as the inaugural Plumian Professor of Astronomy and Experimental Philosophy at Cambridge, while Cotes edited the second edition of Newton's *Principia*. Although Bentley boasted that the observatory contained the 'best instruments in Europe', Molyneux does not add to the debate on their quality; however, Flamsteed, who opposed Cotes's nomination for the Chair, was less complimentary, writing, 'I saw nothing there that might deserve your notice'.[244] Molyneux does refer to the brass sextant by John Rowley and indicates that Newton intended to endow the university, perhaps with the clock which is now in the Master's Lodge at Trinity.

Molyneux clearly approves of Sir Christopher Wren's Trinity College Library, the first significant Classical building in Cambridge, built between 1676 and 1684. He does not review the library collections, but does single out a 'Lucan' manuscript, most likely the *c.* 1406 copy of *Boethius De consolatione philosophiae liber* (still in the Trinity collections today) and a portrait by Kneller of Lord Halifax who was an alumnus of the College of which he was elected a Fellow in 1683.

At this point the manuscript abruptly ends mid-sentence, showing that more missing folios would have followed. Assuming that the journey from

243. Ibid., p. 487.

244. Domenico Bertoloni Meli, 'Roger Cotes (1682–1716)', *Dictionary of National Biography Online* (Oxford, 2004) [accessed 18 June 2010].

Cambridge to Harwich would have taken two days by coach, the date of this final letter suggests that Molyneux spent only two full days in the university town, supporting von Uffenbach's contention that there was little to detain the interested traveller in Cambridge. The letter is despatched from Harwich, from where, on or soon after 20 April 1713, Molyneux departs for Europe. Once abroad, he joins the Electoral Court at Hanover where in June 1714 he witnesses the death of Sophia (1630–1714), the Dowager Electress of Hanover and heir to the English throne, in the garden of her country retreat at Herrenhausen. Molyneux returned to England in June 1714, some five weeks before the death of Queen Anne on 1 August 1714.[245]

Text of Letter 7, one fragment

[f. **155** is blank]

[**156**] D$^{r:}$ Sir Harwich 20 April 1713

Last Wednesday being the 15th of this Month I left London as I told you in my last I was about to do. We were in the Coach by 10 $^{a:m:}$ & passed thro' a very fine wood'd and improv'd Country, full of little Hills and Vales, I remember also we pass'd over the River Lea at a Village call'd Bow in Essex and by several other Villages leaving a very fine seat of Sr Richd Childs call'd Eping Forest to the Right and another of L^{ds} North & Greys on the left and arriv'd in 4h½ at Eping Town. Here we din'd and met much the Country in the afternoon we pass'd by a fine Seat to the Right of [gap] & thro a good handsome Village and in about 3h: came to Hockerhill or Bishop Storford in Harfordshire a good Town or rather two joyn'd. On Thursday the 16 we were much earlier in the Coach I think about 7 a.m. We drove again thro' a mighty fine Country somewhat hilly and of a Chalky Soyl and thro' several pretty Villages and came in about 3h½ to Littlebury in Essex We saw very near us on the Right the famous Palace of Audleyene belonging to L$^{d:}$ Suffolk there is now a great part of it [that formerly stood] taken down and yet it still remains a very large noble old Pile of Building; at Littlebury while dinner was getting we diverted Our Selves with viewing the Whimsical Curiositys [**157**] that are shewn in the house of M$^{r:}$ Winstanley who formerly liv'd there, among a great many Follys there were none that I remember but a couple of Chairs, one of which was plac'd so as on sitting in it by a Trapdoor to sink down into a dark Cellar with Pulleys and Weights to raise it up again, another against a Curtain which cover'd a door that on sitting on the Chair open'd and let it run backwards on four Wheels in the legs fitted to 2 equidistant rails on which it would trundle down into a deep Vally at the bottom of the Garden and could not possibly fall to one side or the other, these two deceits were

245. TCD 750/4/1/1494; Archbishop King confirmed that Molyneux came back from Hanover to London by 28 June 1714.

very well perform'd and comical enough, You see in the Garden a Modell of the old Edyston Lighthouse which was built by this Winstanley but blown down and he himself drown'd in it sometime ago, and there with a Windmill Pump & few other Bundles are all that are to be seen at Littlebury. This place abounds with Saffron. After dinner the Country towards Cambridgeshire began to be more open Champain and unimprov'd and I think not so thickly inhabited We pass'd thro a small Village or 2 and over a place call'd Gogmagogg Hills, this lyes very high and seems by a plain on the top surrounded by three Vallums or Ditches to have been formerly entrench'd and design'd for a place of Strength but whether this be a Roman, Danish, or a Brittish remain **[158]** I cannot say, Camden possibly may inform you, it has now a few Stables built on it for running horses, for which the Sod is very proper; and thus in about 3h½ more we arriv'd at Cambridge generally thro' the most abominable deep Roads that could be but very good ones; As soon as we light at Cambridge I waited on Mr: Cary of Trinity College to whom I had a Letter and was immediately invited by him to sup with them in the Publick Hall which it seems is their manner in all the Colleges here, Wee went first to Trinity Chappel which is really very handsome and well beautifid by the famous Dr: Bentley's means who is now Master or Provost of this College, it is pav'd with Black & White Marble wainscoted with Danzick Oak with a very handsome Altar piece and much loftier and larger than any one in Oxford, we saw also in this College over one of their gates a new Observatory which they are building a fan Octangular Shape each side about 10ft: & 20ft: high and over this again a small round dark tower for the Suns Eclipses &c: in the middle of the Leads that cover it, I observ'd also here fix'd against the Wall a Water Barometer but out of Order, one Mr: Coates is now their Professor of Mathematicks and much esteem'd he has got a new Sextant with some Contrivance of his own to furnish this place but I think there are no other Instruments as yet However they have great Expectations **[159]** from a Benefaction that Sr: Isaac Newton has promis'd them, They have also in this College a pretty good Chimical Laboratory, The Library of this College is what is most worth seeing in it and indeed in all Cambridge there is nothing like it all. The College consists of two noble large Courts of hewn Stone, the innermost of which has this Library for one side, it is rais'd on Pillars of Hewn Stone & has clear under it a most noble Portico to walk in, it is about 200ft. long all the Area of the Library in the Middle flagg'd with black and white Marble about 20ft. broad and 9ft. on each side are taken up with the Classes or Shelves which stand out so far and are about the same Height and distance from one another so that each Class makes 3 sides of a Square all full of Books the Windows being all above and very spacious so as to enlighten the room perfectly well and indeed we esteem'd it as beautiful a Shell and disposition for a Library as could be on each Class they design to have a Bust and have gotten some already as also some Pictures of Benefactors and persons educated there, among whom was my Lord Halifax's Picture, Wee saw here some few Medalls in disorder and some M.S.S. one of Lucan pretty good I believe, there lay also a Quadrant of about 4ft: there unmounted. The Shelves in many places are not full but a Benefaction that has been left them of 50^{l} per ann: [manuscript ends at this point]

APPENDIX

Archbishop William King's Letters to Samuel Molyneux, November 1712 to March 1713, Trinity College, Dublin (TCD) 750/4/1/73–4, 100–2, 109, 120–3, 126.

William King (1650–1729) was, according to the *Oxford Dictionary of National Biography*, 'the single most important Irish protestant churchman of his era'.[246] From 1691 he served as Bishop of Derry until 1703 when he was appointed Archbishop of Dublin, a position he held until his death. He was regarded as being excessively strict and maintained an intellectual reputation that was adequately matched by the stature of his large library. As an early member of the Dublin Philosophical Society, King would have been well acquainted with the work of William and Thomas Molyneux, but it was with the young Samuel Molyneux that he co-authored and published a paper in 1708 on the use of seashells as manure.[247] Archbishop King's friendship with Molyneux appears warm and affectionate, extending far beyond a mutual interest in intellectual pursuits.

Amongst King's personal papers, held in Trinity College Dublin, are copy-books of letters in which his correspondence to Molyneux is recorded. Although the correspondence covers a period of two decades, it is five letters in particular that are significant for our purpose in that they coincide with Molyneux's time in London. These letters, mostly written from Dublin, are dated 25 November 1712 (TCD 750/4/1/73–4), 6 January 1713 (TCD 750/4/1/100–2), 22 January 1713 (TCD 750/4/1/109), 20 February 1713 (TCD 750/4/1/120–3) and 7 March 1713 (TCD 750/4/1/126). They are written in response to Molyneux's letters to King dated 19 November 1712, 18 December 1712, 11 January 1713 and 16 February 1713, none of which are known to have survived. Quotes from the folios in King's copy-books are shown in brackets, while references to Molyneux's letters remain in bold text.

Throughout the letters King provides Molyneux with a paternal display of advice and counsel in-between writing about polite masculine interests. Yet, it is apparent that King is not fully taken into Molyneux's confidence

246. S. J. Connolly, 'William King (1650–1729)', in *Dictionary of National Biography Online* (Oxford, 2008) [accessed 17 September 2010].

247. Lord Archbishop of Dublin and Samuel Molyneux, 'An Account of the Manner of Manuring Lands by Sea-Shells, as Practised in the Counties of Londonderry and Donegall in Ireland. By His Grace the Lord Archbishop of Dublin. Communicated by Samuel Molyneux Esq', *DPST*, 26 (1708), pp. 59–64.

regarding his purpose in the capital. Such secrecy is highlighted in a letter dated 28 June 1714 when Molyneux apologises for not being able to give reasons for his journeys.[248] It is also evident that both parties appear to be particularly guarded when it comes to identifiable detail, most likely fearful that the letters may be intercepted. For example, King, unwavering in his support for a Protestant succession, writes furtively of the Duke of Marlborough in the letter dated 25 November 1712:

> I thank you however for the news you communicate to me of [...] the protestants of Ireland [who] have a great regard for the great man you mention not only on the account of the great actions but also because he seemed to have a great kindness [...] they seem impatient to hear that he is out of Great Britain. (f. 74)

Showing support for the exiled duke was dangerous especially in light of popular thought that Marlborough had prolonged military activities abroad to further his own ambitions and damning contemporary pamphlets such as *The Information against the Duke of Marlborough and his answer* (London, 1712). Perhaps, for the same reasons, there are very few topical issues raised. King repeats 'your story of the Chinese letter' (f. 74), which seems to refer to the controversy that arose in 1705 when the Catholic church questioned Chinese idolatry towards their emperor, and, on the topic of 'the men of Paris', King writes 'sorry to hear [...] that they are all counted Deists' and goes on to regard that such belief in God on purely rational grounds and not through religious texts was an 'absurd doctrine' (f. 121).

Whether King was in Molyneux's confidence prior to his departure for the Hanoverian Court also remains vague. On 6 January 1713 the archbishop remarks on Molyneux's travel plans; he writes '[by] convers[ing] with men of different sentiments [...] the more is to be learned by them and perhaps this is one of the greatest advantages of travelling' (f. 101). However, his comments may well just relate to Molyneux's journey from Dublin to London. King twice stresses the importance of mastering foreign languages. The first reference is briefly mentioned on 6 January 1713 when King writes, 'I believe you will have some use of French and Italian and therefore may practice them all you can in London where I understand you have good opportunities' (f. 100), and again on 20 February 1713, 'You'll reap great advantages by yours having mastered the languages of the countrys where you are to travel' (f. 121). Intriguingly, the 6 January letter implies that Molyneux is considering a difficult decision; King wrote, 'I am of the opinion you may keep your mind to your self and it is a matter for such moment that you ought to take time before you determine' (f. 101) — whether these comments relate to Molyneux's decision to go abroad in the

248. TCD 750/4/1/1494.

service of the duke or to pursue an education is not clear. Two weeks later King surreptitiously offers a word of caution against travelling overseas, quoting Horace, he writes, 'nequiquam deus abscidit prudens oceano dissociabili terras, si tandem inpiae, non tangenda rates transiliunt', which translated reads,

> Heaven's high providence in vain,
> Has sever'd countries with the estranging main,
> If our vessels ne'ertheless,
> With reckless plunge that sacred bar transgress (f. 120)[249]

This could lead to the supposition that King was Molyneux's patron whilst in London. The archbishop certainly would have been well placed to provide contacts and letters of introduction in order to smooth Molyneux's path in the city. He was well associated with 'Mr Churchill' (f. 74) (f. **148**), Dr Woodward (f. 74) (f. **93–7**), Mr Robartes (f. 101) (f. **146**), the various churchmen, all of which were visited by Molyneux, and scientists, such as, 'Dr Isaac Newton', to whom King sends 'my most kindly and heartily respects' (f. 123), as well as asking the question, 'Pray are we to expect a new edition of his *principles*' (f. 123). Yet, there is no firm evidence to suggest any patronage from King and perhaps the idea becomes less likely when, despite the obvious connections, Molyneux does not appear to share his London experiences with him. Indeed, of the notable people he met and the significant places and collections he visited, only Signor Francesco Bianchini features, of whom King writes, 'We have heard of Signore Bianchini and his reception in Oxford, it gives disgust to some, [it has been reported that he is] the Pope's nephew, and it shows the difficulty of his circumstances' (f. 123).

That King took a great interest in Molyneux's personal welfare cannot be doubted. Of getting involved in party politics, King advises, 'I must again and again entreat you not to enter into party's, your youth and design of seeing the world will exclude you from saying anything on such subjects' (f. 74). He also places great emphasis on 'the strongest tys of mutual affection' (f. 100), eulogising the importance of mixing in polite society and waiting on 'persons of quality that take it for an honour to be visited'. Such socialising came with the warning to keep out of 'other men's business', the consequences being 'every one will be looking to introduce you into a learned mans company or a great mans [...] you must be prepared to be judged by persons that have no real kindness for you, and wrong you with insincerity' (f. 121). King's again gives advice by quoting Horace, writing, 'Scaeva; quamvis, scaeva satis per te tibi consulis et scis quo tandem pacto deceat maioribus uti, disce, docendus adhuc quae censet amiculus &c':

249. John Coninton (trans.), *Odes and Carmen Saeculare of Horace* (Oxford, 1892), p. 5 (from Book I Ode III, lines 21–4).

Though instinct tells you, Scaeva, how to act,
And makes you live among the great with tact,
Yet, hear a fellow student; tis as though,
The blind should point you out the way to go,
But still give heed, and see if I produce,
Aught that hereafter you may find of use (f. 100)[250]

Despite the serious advice and support offered by King, the correspondence follows two recurrent themes — books and science.

Molyneux certainly visited bookshops; as a keen collector himself he frequented Bateman's the bookseller in London (f. **151**), and Thomas Hearne associates him with Clement's the bookseller in Oxford. King's letters, however, clearly indicate that Molyneux regularly sourced, bought and sent books to Dublin via King's Chester-based agent, Alderman Massey. There are several references in King's letters to book acquisitions, such as, 'I have one of the *Caesars Commentaries* coming to me [from] Mr Churchill' (f. 74); 'I now thank you for securing Lelands itinerary for so reasonable a price' (f. 109) and 'I wish you could get me the *Reflexions sur eux qui sont morts en riant*' (f. 109). In spite of payment being received through 'Finley' (f. 126), King's London bankers, the letters do not make clear if these transactions were simply a goodwill gesture between two great friends, or whether Molyneux acted as a bibliographical agent for financial gain. King's letter dated 7 March 1713 confirms the business arrangement between the two men, the content referring wholly to dispatching books and arranging payment, but does not elucidate the personal arrangement any further. Later correspondence makes it clear that after Molyneux's departure abroad references to books lessen and a wider agenda of political themes develop.[251]

Within the letters, King discusses 'as a diversion' scientific and mathematical theories in some length and disputes a published mathematical theory concerning globes and spheres of 'Mr Robartes', which 'I read [and] order him my thanks, he did me the honour of a letter on the same subject to which I returned to him my answer' (ff. 101–2).[252] King concludes his line of reasoning with 'I wish I may never have a more material contest with any of my friends especially Mr Robartes whom I greatly value' (f. 102). Francis Robartes may well have been known to both men through his presidency

250. John Coninton (trans.), *The Satires, Epistles and Art of Poetry of Horace* (Oxford, 1892), p. 136 (from Epistulae Book I, 17).

251. King writes to Molyneux on 17 July 1722: 'I rec'd yours of the 3rd. Inst. And the instruments and Homer are come into my hands [with] Dr Flamsteed's book [*Historia Coelestis*] which is very dear indeed but there not being one in Ireland. I am very well Satisfyed and thankfull to you for it'.

252. Francis Robartes, 'Concerning the Proportion of Mathematical Points to Each Other, by the Honourable Francis Robartes Esq. Vice President of the Royal Society', *Philosophical Tranactions*, 27 (1710–12), pp. 470–2.

of the Dublin Philosophical Society in 1693 and through his earlier published research, particularly 'Concerning the distance of fixed stars' (1693) and 'An arithmetical paradox concerning the chances of lotteries' (1694). King also mentions and considers aspects of *Cometographia* by John Hevelius (Danzig, 1668), 'Dr Halleys book on the irregularitys of the moon' (f. 109) and 'a discourse of Dr Halley on comets' (f. 122).

Archbishop King's letters do not overtly help us understand Molyneux's travels in England, nor do they fill in any specific details from the missing folios. They do, however, show that a paternal friendship existed between the senior churchman and the budding philosopher, and that having friends in high places with similar interests certainly eased a stranger's path into London society. The letters also imply that Molyneux was set to go abroad by January 1713, but whether the journey was for business or pleasure we cannot be certain. King's remark, 'you are to be a philosopher' (f. 74), would seem to imply that it was the archbishop's belief that Molyneux's time in London was spent in the honourable pursuit of self-fulfilment. Indeed, just over half a decade earlier the philosopher Gottfried Wilhelm Leibniz wrote, 'This philosophy is a gift from God [...] to serve as the only plank, as it were, which pious and prudent people may use to escape the shipwreck of atheism which now threatens us'.[253] That Molyneux's quest for learning in natural philosophy was honourable cannot be disputed, but these were dishonourable and quarrelsome times, so how far he used his contacts and interests as a subterfuge for political and social ambition still remains unclear.

253. Neil Stephenson, 'Atoms of Cognition: Metaphysics in the Royal Society 1715–2010', in Bill Bryson (ed.), *Seeing Further, the Story of Science and the Royal Society* (London, 2010), p. 84.

EPILOGUE

SAMUEL MOLYNEUX, NATHANIEL ST ANDRÉ AND MARY TOFT

Sheila O'Connell

As Paul Holden notes, the memory of Samuel Molyneux's relatively short life is marred by the unfortunate incident of Mary Toft, the Rabbit Woman of Godalming. Molyneux was drawn into the Toft affair by the man who was to marry his widow: Nathaniel St André, German-speaking surgeon and anatomist to George I who himself spoke no English.[1] St André had come to fame on the wave of the interest in anatomy that was part of the development of empirical science with which Molyneux was so deeply concerned. St André was a serious scientific thinker who contributed to the *Philosophical Transactions of the Royal Society*, but his flamboyant personality led many contemporaries to dismiss him as a charlatan. It was inevitable that when in November 1726 St André heard of a young woman from Godalming who had reportedly given birth to fourteen rabbits, his scientific curiosity coupled with his taste for the sensational led him to hurry down to Surrey to witness the extraordinary deliveries. Samuel Molyneux, a fellow member of the royal court, went with him. Molyneux's interest in 'curiosities' of every kind is evident in the lively descriptions of collections in London and Oxford transcribed in this publication, and he must have been eager to witness this strange phenomenon.

'Monstrous births' had long been a source of fascination whether as portents of doom for the superstitious, or matters of wonder for the curious. Mother and offspring would be displayed at fairs or taverns along with people who were very tall, very small, or very fat, and with exotic beasts and people from far off lands. Shakespeare had Trinculo declare on coming across Caliban: 'A strange fish! Were I in England now, as once I was, and had but this fish painted, not a holiday fool there but would give a piece of silver, There would this monster make a man; any strange beast there makes

1. The Oxford Dictionary of National Biography (2004) includes biographies of William Molyneux, his brother Sir Thomas, and Samuel Molyneux; the Mary Toft affair is, however, mentioned only in the ODNB biography of Nathaniel St André written by Dennis Todd. Todd gives a comprehensive account of the Toft affair in *Imagining Monsters: Miscreations of the Self in Eighteenth-Century England* (Chicago, 1995).

Fig. 1 James Vertue after George Vertue, 'The Surrey Wonder, an Anatomical Farce' *(© Trustees of the British Museum)*

a man; when they will not give a doit to relieve a lame beggar they will lay out ten to see a dead Indian'.[2] Mary Toft, the young woman in question, was evidently hoping to cash in on this taste for the bizarre.

Her husband, Joshua Toft, worked in the cloth trade, a notoriously precarious way of earning a living and one that had notably declined in west Surrey over the previous decades. In the summer of 1726 Mary Toft miscarried her fourth child. What seems to have happened next is that the Tofts, together with Joshua's mother, a midwife, fabricated a monstrous birth by inserting parts of a cat, an eel, and a rabbit into Mary's vagina. A neighbour was called in to witness a 'birth'. This first delivery was followed by a series of rabbits - or parts of rabbits. Mary claimed that early in her pregnancy she had been startled by a rabbit while working in the fields. It was a widely held belief - even in medical circles - that experiences having

2. William Shakespeare, *The Tempest*, Act II, scene ii.

Fig. 2 William Hogarth, 'Cunicularii, or the Wise Men of Godliman in Consultation'
(© Trustees of the British Museum)

a powerful effect on a mother's imagination during pregnancy could affect the unborn child, for instance, a mother frightened by a hare might give birth to a child with a hare lip or cleft palate.

John Howard, a local surgeon and man-midwife, was soon involved, but it is not clear whether or not he was part of the conspiracy. At the beginning of November he brought the Tofts to lodge near his home in Guildford where rabbits – always dead – continued to be delivered. Howard preserved the specimens and made detailed notes. News spread rapidly of 'the rabbit woman'. St André and Molyneux arrived in Guildford on 15 November. St André delivered part of a fifteenth rabbit which appeared in pieces over several hours that day. Molyneux certainly examined Mary Toft and may also have assisted with the delivery - both Lord Hervey and the astronomer and theologian William Whiston recalled him claiming very firmly that he had done so.[3] Molyneux and St André studied the creature and the earlier

3. John Hervey, Baron Hervey, letter to Henry Fox, 3 December 1726, in Giles Stephen Holland Fox-Strangways, Earl of Ilchester (ed.), *Lord Hervey and His Friends 1726-38* (London, 1950); William Whiston, *Memoirs of the life and writings of Mr William Whiston: containing memoirs of several of his friends also* (London, 1750), Part III, p. 111 (both quoted by Todd, pp. 40-41).

Fig. 3 William Hogarth 'Credulity, Superstition and Fanaticism'
(© Trustees of the British Museum)

specimens with great attention, and were convinced of their authenticity. Back in London St André put the case to George I who was intrigued and sent him back to Guildford with two eminent man-midwives, Sir Richard Manningham and Philip van Limborch.

Manningham and Limborch brought Mary Toft up to London on 29 November and lodged her in Lacy's bagnio in Leicester Fields where according to Lord Hervey, 'Every Creature in town both Men & Women have been to see & feel her'.[4] She appeared to go into labour on several occasions, but there were no further deliveries. A succession of doctors, as well as representatives of the law and other prominent gentlemen questioned her closely, and then a porter at the bagnio was found with a rabbit in his pocket. Meanwhile Lord Onslow, Lord Lieutenant of Surrey, had taken depositions from a number of local people who had sold rabbits to Joshua Toft over the past month. Manningham, who had at least at first, been intrigued by the possibility that monstrous births were actually taking place, appealed to Molyneux for help in keeping Mary out of prison long enough to find out exactly what had been going on.[5] He ended by threatening her with a painful procedure to see if she had some anatomical peculiarity - and finally, on 7 December, she confessed.

Those who had been duped were held up to ridicule in the newspapers, and in pamphlets, songs, and satirical prints: 'The wonder of wonders ...', 'the wonderful coney-warren, lately discovered at Godalmin ...', 'A New Whim-Wham from Guildford', 'The Anatomist Dissected, or, The Man-Midwife Finely Brought to Bed', 'St André's Miscarriage' (in which Molyneux is referred to as 'Molly'[6]), 'The Discovery, or the Squire turn'd Ferret' by Alexander Pope and William Pulteney (see Appendix).[7] A bawdy satire, 'Much Ado about Nothing: or, the Rabbit-Woman's Confession', supposed to have been written - with atrocious misspelling - by Mary Toft herself, identifies Molyneux by reference to his interest in telescopes.[8] She records that one of those who came to visit her was a 'purblynd Gentilman hoo was for survayin me with his telluskop; but it was so dark he cood not see, tho he got upon a gynt-stool and had it not bin for anothur Parson more quick-sited then himself, he had sartinly lost his Telluskop'.

4. Hervey, op.cit.
5. Richard Manningham, *An Exact Diary of what was observ'd during a close attendance upon M. Toft, the pretended Rabbet-Breeder* (London 1726).
6. 'Molly' was an offensive term for a homosexual.
7. There are two contemporary collections of pamphlets on the Toft affair in the British Library, pressmarks 1178.h.4 and G.1394.
8. Merry Tuft, *Much Ado about Nothing; or, a plain refutation of all that has been written or said concerning the Rabbit-woman of Godalming, being a full confession from her own mouth, and under her own hand, of the whole affair* (London, 1727).

In a print by James Vertue called 'The Surrey Wonder, an Anatomical Farce'[9] (Fig. 1) Molyneux is portrayed raising his hands in disgust as a woman holds up a rabbit to view. The print was inspired by a 'rabbit scene' introduced into 'The Necromancer', a popular pantomime at the Lincoln's Inn Theatre. The most accomplished print on the subject was William Hogarth's 'Cunicularii, or the Wise Men of Godliman in Consultation'[10] (Fig. 2) which, according to Hogarth's early biographer John Nichols, was commissioned by a group of rival surgeons paying a guinea each.[11] Mary Toft is shown in the throes of labour attended by Joshua Toft and his sister Margaret, who acted as a nurse, as well as by John Howard who is receiving a dead rabbit at the door. There are three other men in the room (fewer than most accounts record): St André dances in excitement proclaiming, 'A Great Birth'; John Maubray, whose recent book on the birth of 'sooterkins', small furry creatures, to mothers in the Netherlands gave some medical credence to Mary Toft's story;[12] and a third whose arm is thrust beneath Mary's skirts. This last is usually identified as Manningham, but Dennis Todd makes a case for identifying him as Molyneux, or at least a conflation of Molyneux and Manningham.[13] Hogarth's key describes him as 'An Occult Philosopher searching into the Depth of things', a phrase which seems to refer to Molyneux's interest in telescopes and would have appealed to those who regarded scientific pursuits as tainted by elements of both magic and fraud. Hogarth was to revive memories of Mary Toft more than thirty years later when he introduced her, with small rabbits running out from under her skirts, as an illustration of credulity in his print 'Credulity, Superstition and Fanaticism' (Fig. 3), published in 1762.[14]

While the Mary Toft affair ended in nothing more than mockery and embarassment for those who had believed her story, Molyneux and St André were brought together in the public domain just sixteen months later in circumstances that can only be described as tragic. Molyneux died suddenly and St André was accused of poisoning him.

Early in 1728 Molyneux collapsed in fit while in the House of Commons. He died a few days later at three o'clock in the morning of 13 April at his house in Great George Street[15] near Hanover Square. His friend St André

9. Frederic George Stephens, *Catalogue of Political and Personal Satires in the British Museum*, vol. II, pp. 633-38, no. 1778.
10. ibid., vol. II, pp. 638-39, no. 1779; Ronald Paulson, *Hogarth's Graphic Works* (1989), pp. 69-70, no. 106.
11. John Nichols, *The Genuine Works of William Hogarth* (1808-10), vol. I, pp. 37-8, vol. II, pp. 49-60.
12. John Maubray, *The Female Physician, containing all the diseases incident to that sex*, (1724).
13. Todd, op.cit, pp. 92-94.
14. Paulson, *op.cit.*, pp. 175-78, no. 210.
15. now St George Street.

had been treating him, and on the night of Molyneux's death eloped with his widow, Lady Elizabeth. Society was horrified at this indiscretion and the couple lost their places at court. They married in 1730 and settled in Hampshire. By the time of Lady Elizabeth's death in 1759, she was clearly on good terms with her first husband's family: her will lists a considerable estate, most of which was left to St André, but she bequeathed to Sir Capel Molyneux "my best ruby formerly set in a ring for my late husband ... [and] all such plate as shall be found to have only the arms or crest of the said Samuel Molyneux engraved on it".[16] Capel was Lady Elizabeth's namesake, born in the year of her marriage to Samuel; his father was Sir Thomas Molyneux to whom the letters in this publication were written.

The Discovery, or the Squire turn'd Ferret

by Alexander Pope and William Pulteney

Most true it is, I dare to say,
E'er since the Days of Eve,
The weakest Woman sometimes may
The wisest Man deceive.

For D[avena]nt[17] circumspect, sedate,
A Machiavel by Trade,
Arriv'd Express, with News of Weight,
And thus, at Court, he said.

At Godliman, hard by the Bull,
A Woman, long thought barren,
Bears Rabbits, - Gad! so plentiful,
You'd take her for a Warren.

These eyes, quoth He, beheld them clear:
What, do ye doubt my View?
Behold this Narrative that's here:
Why, Zounds! and Blood! 'tis true!

Some said that D[ou]gl[a]s sent should be
Some talk'd of W[a]lk[e]r's merit,[18]
But most held, in this Midwifery,
No Doctor like a FERRET.

But M[o]l[y]n[eu]x, who heard this told,
(Right wary He and wise)
Cry'd sagely, 'Tis not safe, I hold,
To trust to D[avena]nt's Eyes.

16. Will of Lady Elizabeth St André, TNA PROB 11/845.
17. Henry Davenant (b. 1679), a well-placed diplomat and courtier to George I, visited Mary Toft in Guildford early in November and returned to London evidently convinced that she had indeed produced rabbits.
18. Two well-known man-midwives, James Douglas (1675-1742), who was present at Mary Toft's confession, and Middleton Walker (d. 1732).

A Vow to God he then did make
He would himself go down,
St. A[n]dre too, the Scale to take
Of that Phoenomenon.

He order'd then his Coach and Four;
(The Coach was quickly got 'em)
Resolv'd this Secret to explore,
And search it to the Bottom.

At Godliman they now arrive,
For Haste they made exceeding;
As Courtiers should, whene'er they strive
To be inform'd of Breeding.

The good Wife to the Surgeon sent,
And said to him, Good Neighbour,
'Tis pity that two Squires so Gent-
Should come and lose their Labour.

The Surgeon with the Rabbit came,
But first in Pieces cut it;
Then slyly thrust it up that same,
As far as Man could put it.

(Ye Guildford Inn-Keepers take heed
You dress not such a Rabbit,
Ye Poult'rers eke, destroy the Breed,
'Tis so unsav'ry a-Bit.)

But hold! says Molly, first let's try,
Now that her Legs are ope,
If ought within we may descry
By Help of Telescope.

The Instrument himself did make,
He rais'd and level'd right,
But all about was so opake,
It could not aid his Sight.

On Tiptoe then the Squire he stood,
(But first He gave Her Money)
Then reach'd as high as e'er He could,
And cry'd, I feel a CONY.

Is it alive? St. A[n]dre cry'd:
It is; I feel it stir.
Is it full grown? The Squire reply'd,
It is; see here's the FUR.

And now two Legs St. A[n]dre got,
And then came two Legs more;
Now fell the Head to Molly's Lot,
And so the Work was o'er.

The Woman, thus being brought to Bed,
Said, to reward your Pains,
St. A[n]dre shall dissect the Head,
And thou shalt have the Brains.

He lap'd it in a Linnen Rag
Then thank'd Her for Her Kindness
And cram'd it in the Velvet Bag
That serves his R[oya]l H[ighness].

That Bag – which Jenny, wanton slut,
First brought to foul Disgrace;
Stealing the Papers thence she put
Veal-cutlets in their Place.

O! happy would it be, I ween,
Could they these Rabbits smother;
Molly had ne'er a Midwife been,
Nor she a shameful Mother.

Why has the Proverb falsely said
Better two Heads than one;
Could Molly hide this Rabbit's Head,
He still might shew his own.

SAMUEL MOLYNEUX'S LONDON

These maps are intended to give some idea of the city in which Samuel Molyneux spent the winter months of 1712–13. It was a small place compared with the London of today. Molyneux had taken rooms in Suffolk Street just to the west of the Royal Mews, today's Trafalgar Square, and a healthy young man could easily have walked to any of the destinations mentioned in little more than half an hour.

We have used John Roque's 1746 *Plan of the Cities of London and Westminster* as a base for the maps and have numbered on it the places mentioned by Molyneux more or less in the order in which he writes of them. When the exact location is uncertain, we have marked the street — for example, we do not know in which house Molyneux rented rooms for his sojourn, only that it was Suffolk Street.

Since Molyneux's visit preceded Rocque's survey by thirty-three years, some of the places that he visited had already been demolished before the map was made. For example, Berkeley House and gardens had given way to Burlington House and the site of Sunderland House had been redeveloped as Sackville Street.

MAP I

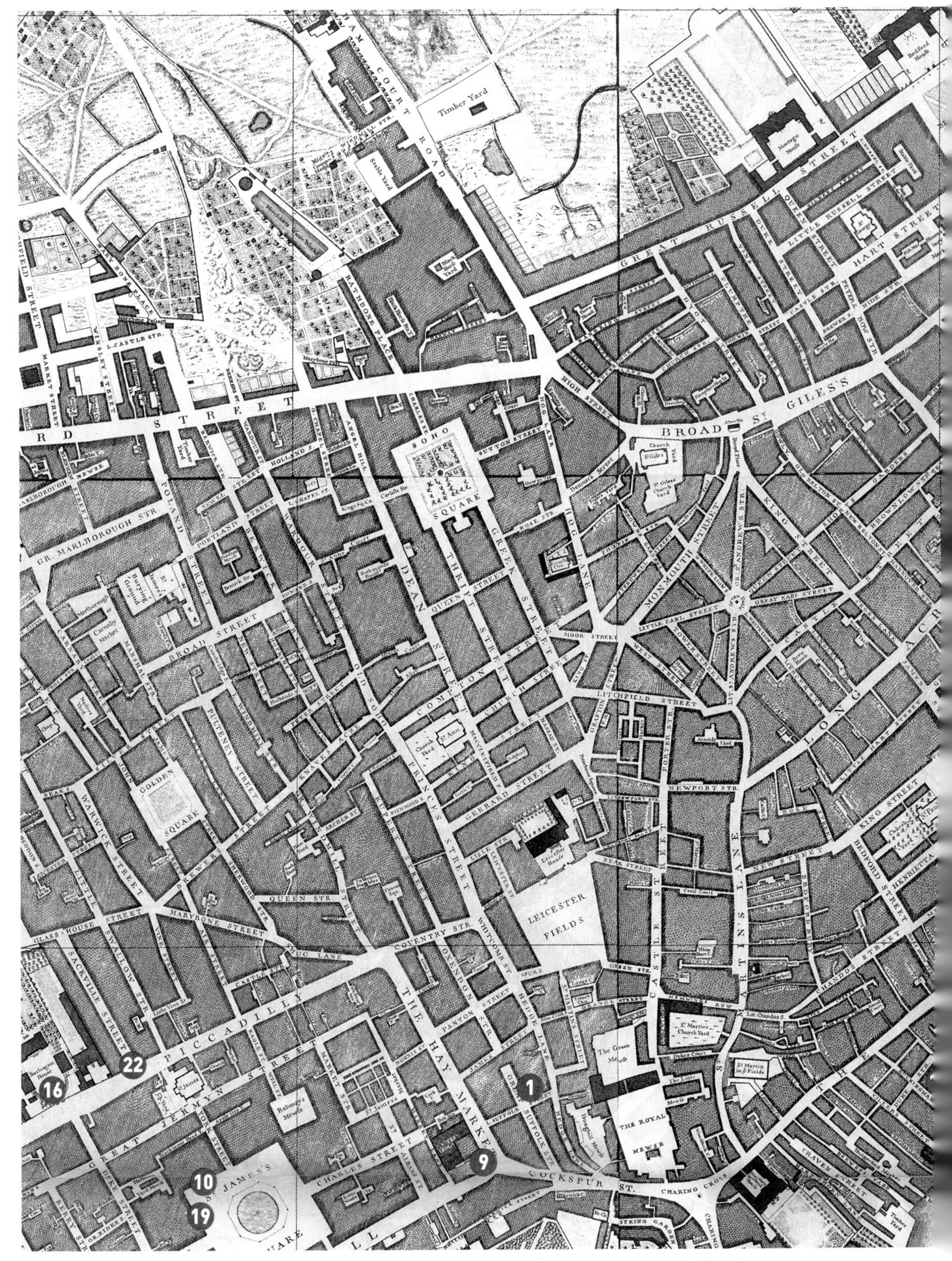

1. Suffolk Street where Molyneux took lodging
9. John Kempe's Collection
10. Lord Pembroke's Collection
16. Berkeley House (site of)
18. Harleian Collection
19. Sir John Germaine's Collection
22. Sunderland House (site of)

MAP II

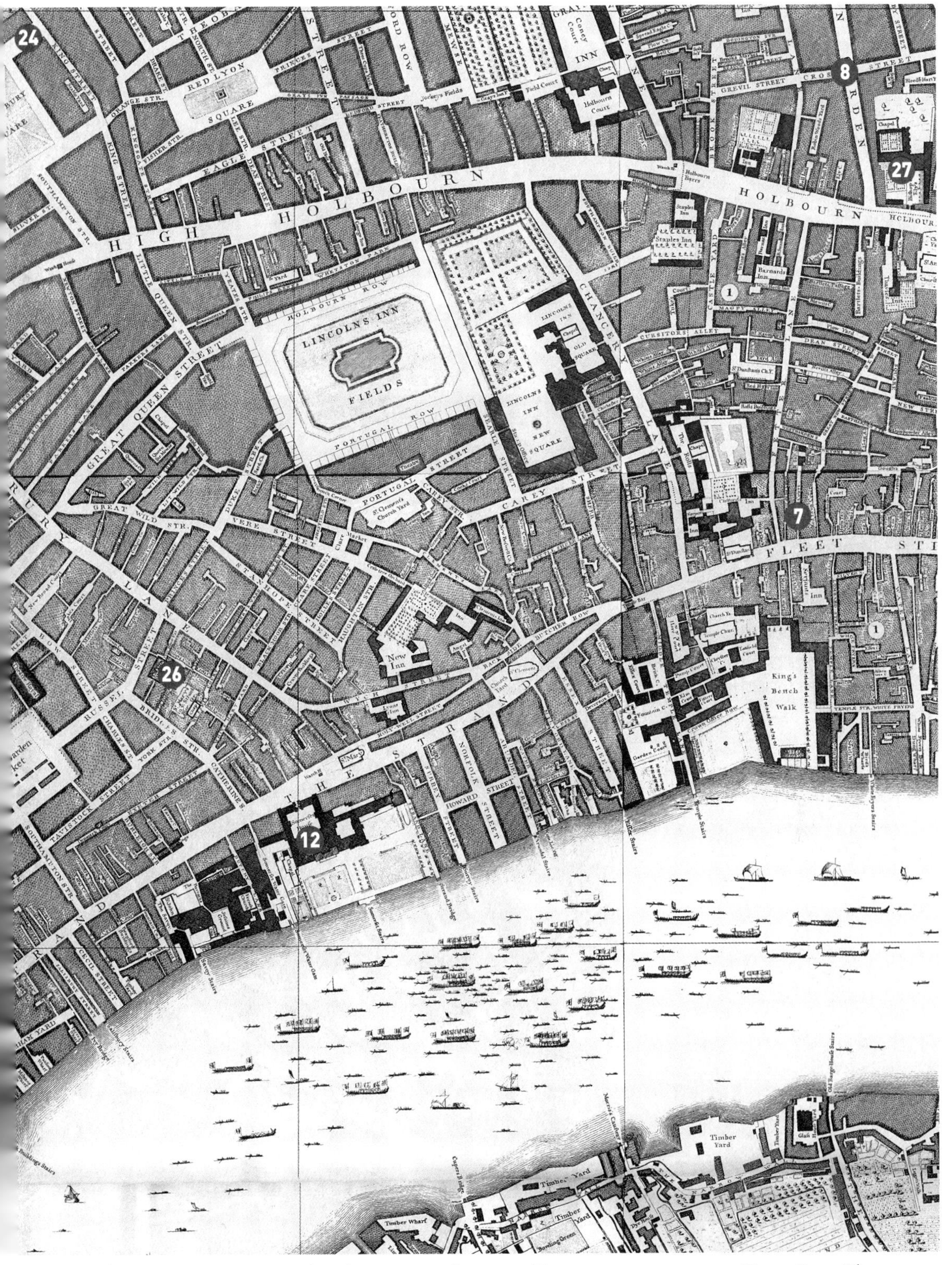

7. Royal Society, Crane Court, Fleet Street
8. James Wilson, astronomical instruments, Cross Street, Hatton Garden
12. Somerset House
24. Sir Hans Sloane's Collection, 8 Bloomsbury Street
26. Drury Lane Theatre
27. Bishop Moore's Library, Ely Place

MAP III

2. St Paul's Cathedral 21. James Petiver's Collection 25. Bateman's Bookshop

MAP IV

6. Tower of London 17. Dr Woodward's Collection, Gresham College

MAP V

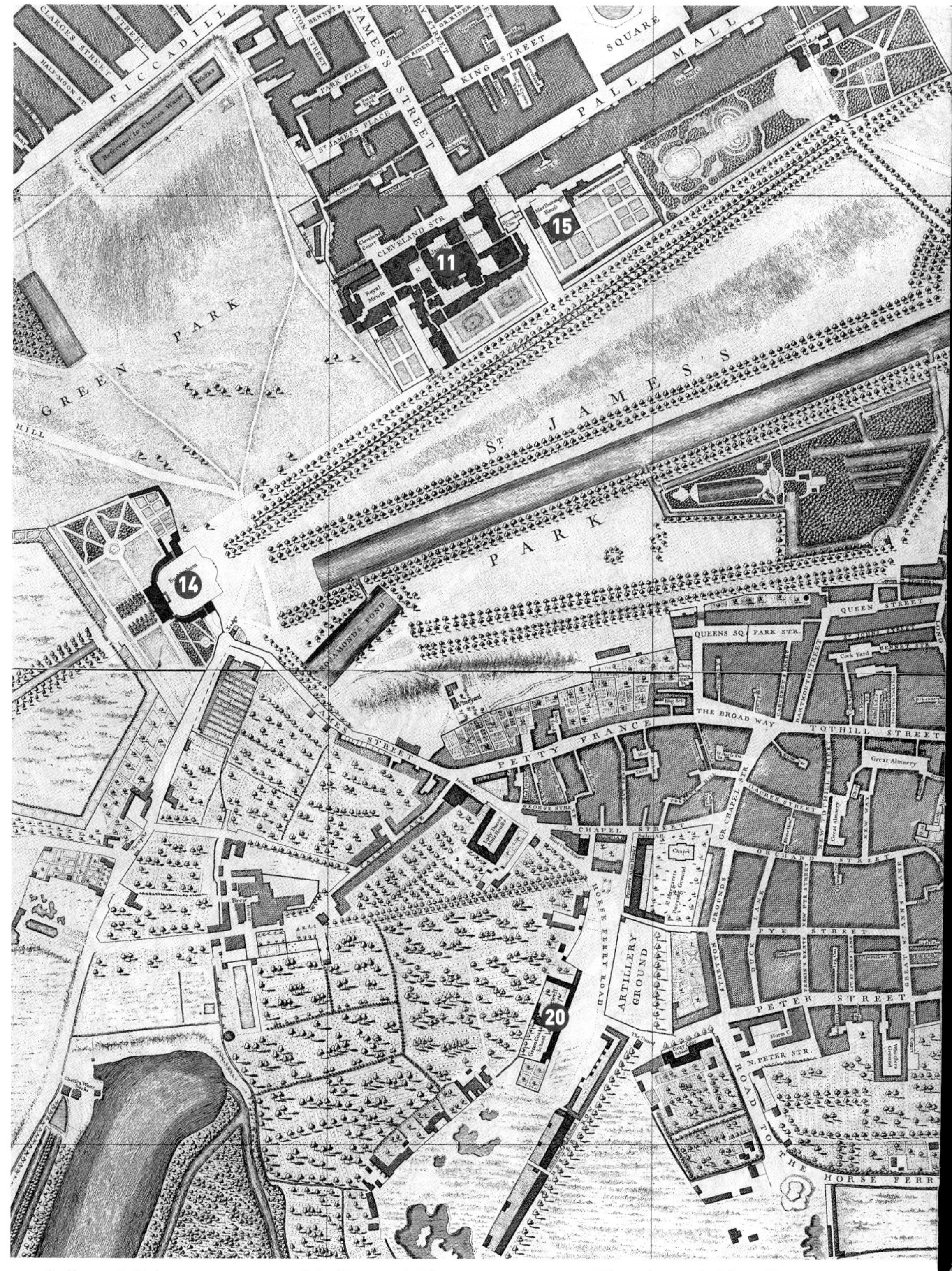

11. St James's Palace
14. Buckingham House
15. Marlborough House
20. Westminster Bridewell

MAP VI

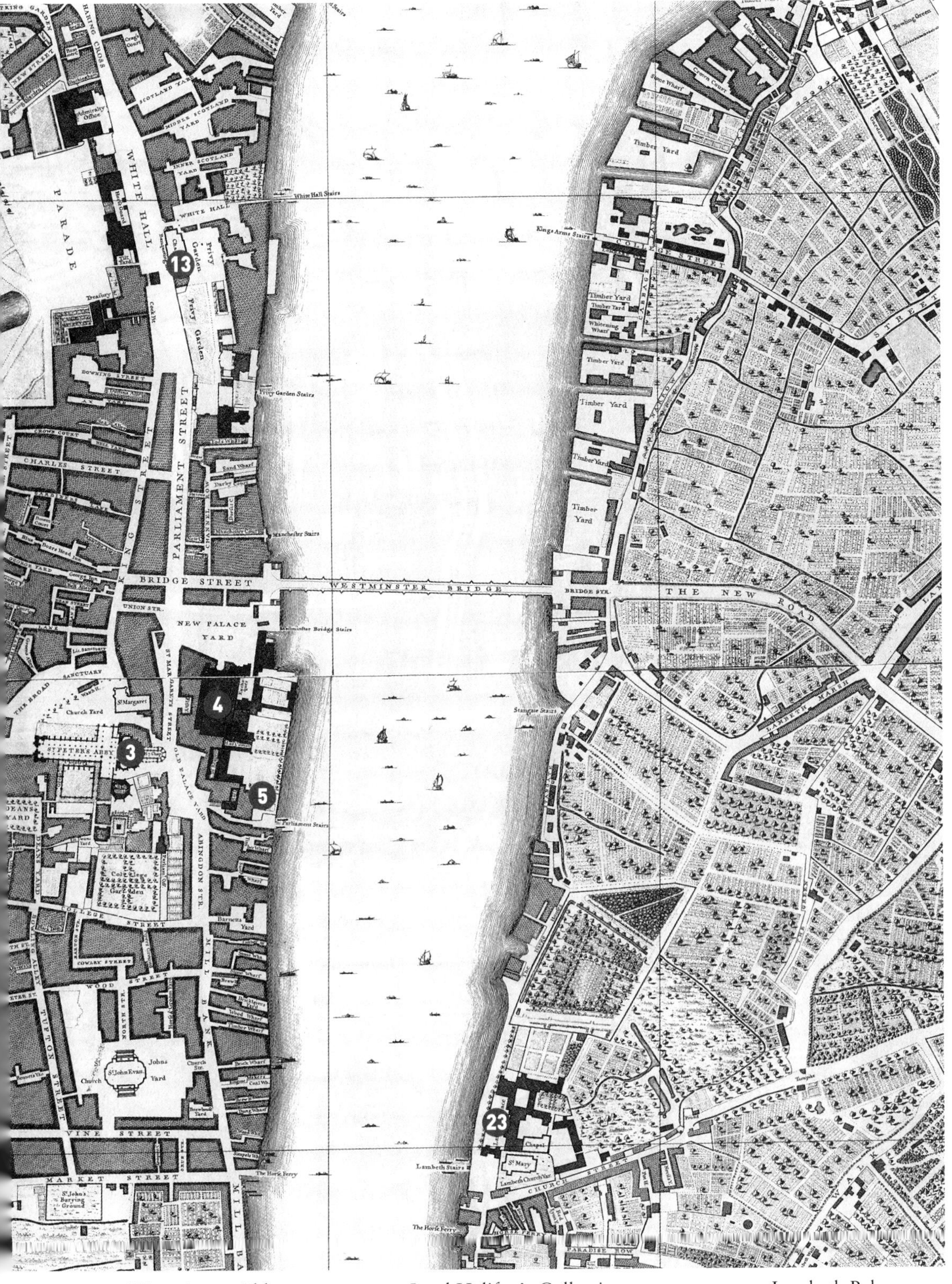

3. Westminster Abbey
4. Westminster Hall
5. Lord Halifax's Collection
13. Whitehall Banqueting House
23. Lambeth Palace

INDEX